Building and Maintaining Award-Winning ACS Student Member Chapters
Volume 1: Holistic Viewpoints

ACS SYMPOSIUM SERIES 1229

Building and Maintaining Award-Winning ACS Student Member Chapters Volume 1: Holistic Viewpoints

Matthew J. Mio, Editor
University of Detroit Mercy
Detroit, Michigan

Mark A. Benvenuto, Editor
University of Detroit Mercy
Detroit, Michigan

Sponsored by the
ACS Society Committee on Education

American Chemical Society, Washington, DC

Distributed in print by Oxford University Press

Library of Congress Cataloging-in-Publication Data

Names: Mio, Matthew J. (Matthew John), 1974- editor. | Benvenuto, Mark A.
   (Mark Anthony), editor. | American Chemical Society. Society Committee on
   Education.
Title: Building and maintaining award-winning ACS student member chapters volume 1 :
   holistic viewpoints /
   Matthew J. Mio, editor, University of Detroit Mercy, Detroit, Michigan,
   Mark A. Benvenuto, editor, University of Detroit Mercy, Detroit, Michigan
   ; sponsored by the ACS Society Committee on Education.
Description: Washington, DC : American Chemical Society, [2016] | Series: ACS
   symposium series ; 1229, 1230 | Includes bibliographical references and
   index.
Identifiers: LCCN 2016043973 (print) | LCCN 2016044818 (ebook) | ISBN
   9780841231696 (alk. paper : v. 1) | ISBN 9780841231719 (alk. paper : v. 2)
   | ISBN 9780841231634 ()
Subjects: LCSH: American Chemical Society--Membership. | Chemistry--Study and
   teaching (Higher)
Classification: LCC QD1 .B81495 (print) | LCC QD1 (ebook) | DDC
   540.6/073--dc23
LC record available at https://lccn.loc.gov/2016043973

The paper used in this publication meets the minimum requirements of American National Standard for Information Sciences—Permanence of Paper for Printed Library Materials, ANSI Z39.48n1984.

Foreword

The ACS Symposium Series was first published in 1974 to provide a mechanism for publishing symposia quickly in book form. The purpose of the series is to publish timely, comprehensive books developed from the ACS sponsored symposia based on current scientific research. Occasionally, books are developed from symposia sponsored by other organizations when the topic is of keen interest to the chemistry audience.

Before agreeing to publish a book, the proposed table of contents is reviewed for appropriate and comprehensive coverage and for interest to the audience. Some papers may be excluded to better focus the book; others may be added to provide comprehensiveness. When appropriate, overview or introductory chapters are added. Drafts of chapters are peer-reviewed prior to final acceptance or rejection, and manuscripts are prepared in camera-ready format.

As a rule, only original research papers and original review papers are included in the volumes. Verbatim reproductions of previous published papers are not accepted.

ACS Books Department

Contents

Indexes

Preface

"It's such an ancient pitch
But one I wouldn't switch
'Cause there's no nicer witch than you
'Cause it's witchcraft
That crazy witchcraft..."
Music by CY COLEMAN, lyrics by CAROLYN LEIGH

Chemists don't often quote a song made famous by Frank Sinatra. Chemists definitely don't often speak of witchcraft—or at least not with a straight face. Yet there remains a certain magic, a wide-eyed wonder, a certain hard-to-define personal "chemistry" when it comes to our love of our science and our deep desire to pass it on to the next generation of student-scientists.

Part of the love of chemistry manifests itself in the hard work that goes into building and then to maintaining a truly first-class American Chemical Society Student Member Chapter. The ingredients are not always the same from one chapter to another, but there is often overlap. Is it the advisor? Is it the student members? Is it the opportunities? Is it all of the above? The "all of the above" response is the most obvious, but there is definitely more to it than putting the right people together in the right place at the right time.

This twin set of volumes is an attempt to capture numerous voices among those in the ACS who have built award-winning Student Member Chapters, as well as those who have kept them going, in some cases for decades. There is a great deal of energy within our ranks. We have tried to capture the experiences of these, some of our most active Student Member Chapter leaders, so that energy can easily be spread to others.

Several of the chapters in these volumes have been written by Student Member Chapter advisors. These individuals are those who may have the longest effect on a Chapter simply because an advisor is usually a faculty member who has built her or his career at an institution. When they have a positive attitude, drive, and energy, and when they bring these to their "Chemistry Clubs," the influence they can have on student members is far-reaching and long lasting.

While faculty members who are deeply involved in ACS Student Member Chapters provide long-term stability and continuity, involved students and active student leaders bring the energy and vitality to Clubs that renews itself annually. Students come to colleges and universities with high hopes and great dreams, and active, enthusiastic "Chem Clubs" can fulfill them. Students also bring new ideas with them, ideas that can be tried, then used or discarded depending on the results.

As students mature in their time in higher education, they can and often do take on active leadership roles and bring a Club's activities and health to new heights.

The energy and hard work of faculty advisors and student leaders doesn't go too far if there isn't a larger student body with which to work, and to whom the excitement and fun of chemistry can be imparted. Quite a few of the chapters in these volumes address at length how their departments, their college or university, or the general public in their area are able to interact with the Student Member Chapter. It is impressive to see what is being done throughout the Society to get some of the joy and fun of chemistry to a large, varied audience.

We have divided the chapters of this project into two volumes, with the split aligning as follows: this volume in which the authors look at their Chapter overall, and discuss all of the facets that make up their organization (Volume 1: Holistic Viewpoints) and the companion volume in which the authors have focused on one or more strengths of their Student Member Chapter (Volume 2: Specific Program Areas). It is hoped that this will give readers some convenient places to start should they desire to see what some of the other shakers and movers are doing throughout the ACS.

We hope these two volumes will give an interested reader many different recipes to build a great Student Member Chapter, and ways to maintain it for years. And even though we think these books provide several excellent, specific recipes for advisors, student leaders, and others interested in making the study and understanding of chemistry something that will be a central part of their undergraduate experience, we also hope there might still be a bit of "that crazy witchcraft" floating about which makes our science so much fun.

Matthew J. Mio
University of Detroit Mercy

Mark A. Benvenuto
University of Detroit Mercy

Chapter 1

Building an Outstanding ACS Student Chapter

Susan A. White, Joseph A. Piechocki, Jr., and Matthew J. Allen*

Department of Chemistry, Wayne State University, 5101 Cass Avenue, Detroit, Michigan 48202
***E-mail: mallen@chem.wayne.edu**

The Wayne State University American Chemical Society-Student Affiliates (WSU ACS-SA) was reactivated in August 2011, after a long absence. The chapter has established a strong presence on campus and has designed its own annual events. The chapter has built relationships within the WSU Chemistry Department, the ACS Detroit Local Section, and other ACS Student Chapters. The WSU ACS-SA was proud to earn the American Chemical Society's Outstanding Student Chapter Award in its fourth year and is working hard in its fifth year to continue, and build, on its successful traditions. The WSU ACS-SA is striving to create a legacy that will endure long after the current members graduate so that WSU will continue to have a strong ACS student presence on campus. The lessons learned by the students and their advisors during the journey from nothing to a thriving chapter will likely be of help to universities looking to begin active chapters. The approach taken by the WSU ACS-SA is described in this chapter.

Introduction

Promoting chemistry and chemical education is a challenging task. One way to encourage students to become involved is by seeing how much other interested students enjoy learning and promoting the subject. Establishing a student chapter of the ACS is a great way of engaging a student population to boost interest in chemistry. However, a chapter is unlikely to begin without determined and passionate students, and the process can appear so daunting that many students will not want to participate. In the course of starting a chapter, several questions need to be asked: How does a chemistry club start? What is needed to build a student chapter? And, how will the chapter continue to be successful each year? The re-establishment of the Wayne State University American Chemical Society-Student Affiliates (WSU ACS-SA) (*1*) chapter is a good example of how a club can quickly develop and to become positively recognized and supported by chemistry faculty members and university students.

Setting up a chapter to be successful is a challenge; there is not one right way for a chapter to become established, grow, and have members benefit from the professional rewards of the chapter. However, from experience building and growing a chapter over the past five years, the WSU ACS-SA has been able to accomplish many goals, including developing a professional rapport with the Department of Chemistry, networking with other chemistry clubs and the local section, and encouraging students to take an interest in chemistry. In the remainder of this chapter, the route that the WSU chapter took to become a successful student chapter is described along with suggestions to assist other new or reactivated chapters.

Historical Background

The WSU ACS-SA was originally established in 1943, with fifteen ACS members. Since that time, the chapter has had many incarnations. The current chapter was reactivated with the ACS in 2011, when a small group of six dedicated and enthusiastic students approached Professors Matthew Allen and Mary Kay Pflum with the idea to restore the chapter for the 2011–2012 academic year. Under the leadership of the chapter President, Milad Karim, and Vice President, Zeinab Moubadder, the new chapter met all of the requirements of the ACS and the WSU Dean of Students Office in time for the Fall Term of 2011. The leadership of the newly activated club was recognized for their hard work with the ACS's Honorable Mention Award in their first year. The chapter continued to grow, expand, and prosper, and by the fourth year, it earned the ACS Outstanding Student Chapter Award. Since its reincarnation in 2011, the WSU ACS-SA has continued to grow. The chapter currently has approximately 60 paid members and endeavors to be a healthy organization that proudly represents the Department of Chemistry and WSU.

Starting a Chapter

Once the decision to begin or reactivate an ACS student chapter is made, steps can be taken to make the chapter officially recognized by organizations that can help the chapter thrive. There are two ways in which the student chapter can be recognized: as a student organization at the university and as an official ACS chapter. We recommend that new or reactivated student chapters concurrently evaluate what is necessary at both the university and the ACS level because there likely will be overlapping requirements. Being recognized at both levels can be beneficial. At the university, the chapter gains exposure to students and enables networking opportunities with other student organizations. The ACS offers several benefits including access to the network of chemists and chapter development resources offered only for student chapters. Additionally, both the university and ACS can be a source of funds to support activities. To be recognized by the university, the WSU chapter is managed within the Dean of Students Office that provides guidelines and rules that must be followed within organizations or student clubs at WSU. WSU requires that organizations have a club name, a president or leader, and at least one other member. The president and other designated leaders of the organization make up the initial board of the organization. The Dean of Students Office also recommends that an advisor be selected to guide the students and assist with building the chapter. Furthermore, it is required that a constitution be submitted to the Dean of Students Office, where it is maintained while the chapter is active.

Similarly, the ACS chapter guidelines need to be consulted for the rules and requirements for clubs (2). To be recognized by the ACS, a chapter must have at least six paid ACS student members, a faculty advisor, bylaws, and a submitted charter application. The bylaws can also be part of the constitution setup with the college or university, and they are likely to contain some of the same sections mentioned above (Figure 1).

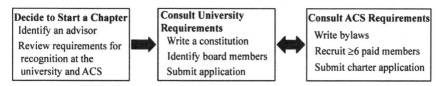

Figure 1. Flowchart of the steps necessary for a club to be recognized at WSU and by the ACS

Two elements that are common between the WSU and ACS are the need for an advisor and a constitution or bylaws. Both selecting an advisor and creating a constitution are crucial to a club's success. In choosing a chapter advisor, the students in the club need to select an advisor with whom they feel comfortable working, who can help make crucial decisions, and who can help guide the chapter. In the WSU chapter, the chapter advisors are involved with some of the activities of the club; however, the advisors do not make final decisions. The WSU ACS-SA chapter is student led, and students are the decision-makers and generate professional alliances and connections. The main roles of the advisors are to provide guidance toward making critical decisions and to help the club navigate within the university structure to achieve the student chapter's goals. In our opinion, the ownership taken by the students is a major reason why the chapter has been successful. Students are able to develop critical thinking skills, professional networking skills, and are exposed to writing and communicating by proposing grants and planning and executing events.

An organized club is difficult to manage without written rules and procedures, and a constitution helps the club maintain an organizational structure. Colleges or universities sometimes have templates for constitutions for student organizations; however, a club likely is not restricted to what the template contains. Some things the constitution at WSU contains are the goals and purpose of the organization, membership requirements, officer definitions and responsibilities, requirements for being an officer, election procedures, defining quorum and methods for holding meetings, and procedures for amending the constitution. Although this is not an all-inclusive list of what a constitution could contain, it includes the basic foundations for beginning the chapter. Once all of the requirements have been met and approved, the student chapter is officially recognized at both the college or university and the ACS.

Creating a Strong Foundation

Establishing Regular Meetings

Once the executive board members are selected, it is important to decide the exact details of the club's member and executive meetings. Based on the experience of the WSU ACS-SA, meetings are best attended if scheduled to occur regularly at the same time and place. Chapters can choose a frequency (weekly, bi-weekly, or monthly) that suits the chapter, but setting the schedule at the beginning of each semester offers predictability for busy students. At WSU, attendance of member meetings is the largest and most consistent throughout the semester when the time and place are announced either early in the semester or before the semester begins. Attendance is also largest when the meeting schedule does not change. A meeting time strategically determined by the board members maximizes the availability of students to attend. The WSU ACS-SA chapter tries to avoid the times of all of the major chemistry classes and picks a time attractive to students when the largest number of board members can attend. At WSU, early morning meetings or meetings on Friday afternoons have not worked well, so those times are avoided. The WSU ACS-SA constitution requires that board members attend the meetings; as a result, meetings are scheduled to avoid the classes and obligations of all board members when possible. We recommend that meeting times and places are set before a chapter starts to heavily recruit members. Selection of a schedule prior to widespread recruiting permits advertisement of the meeting times on recruiting flyers, social media, or websites. In addition to member meetings, regular executive board meetings have been vital for club management at WSU. At the board meetings, issues for the next member meeting are discussed and all events and activities are voted on and arranged. The time spent in board meetings yields shorter, more engaging, and better prepared member meetings. We find that weekly meetings for both the board and the members allow the club to keep everyone informed and add structure to the administration of the club.

At the initial member meeting of a semester at WSU, the club executive presents a brief introduction of the board, an overview of the club, a schedule for the semester, and an explanation of all the opportunities offered by the chapter and the ACS. It is important to describe the offered opportunities at the first meeting because students want to know if investing in the club will be worth the benefits and skills that they can obtain. In addition to the above presentation, the organization distributes informative flyers and registration forms that potential members can complete. For WSU, the best meeting attendance occurs when coupled with small promotions. Free food and small giveaways, like swag from the ACS National Meeting, help increase attendance. The best meeting attendance for 2015 involved liquid nitrogen safety training, required for making ice cream, combined with sampling the product.

Developing and Arranging Events

At WSU, any discussion about scheduling meetings for the next year is generally accompanied by a more complete annual agenda including the potential chapter events. The ACS student magazine *inChemistry* often highlights many of the original, fun, and interesting events held by student chapters around the country. The choices for possible events are nearly endless, but the WSU ACS-SA has concentrated on two large annual events along with a multitude of smaller activities. The WSU ACS-SA annual events are the Faculty–Student Mixer and the For the Love of Chemistry Symposium. The Faculty–Student Mixer is held in the atrium of the Chemistry Department in November. The mixer encourages students to meet and speak with chemistry professors and fellow students. After the professors are introduced, the students and faculty mingle, providing a low-pressure opportunity to ask questions. Students can discuss possible majors, courses, and research options with the professors and other students. Numerous undergraduates have found a research position through participation in this mixer. The second large annual event is our multi-school For the Love of Chemistry Symposium traditionally held in February. This has been the WSU ACS-SA's largest event each year. Attendance at this event has averaged over 100 people per year for the past three years, and in 2016, it was attended by students from seven schools. This event was inspired by our chapter's love of chemistry and our desire to interact with other chapters. This all-day event starts with four career speakers, each representing a different chemistry-related field. After the career talks, the speakers form a panel to answer questions about their careers and chemistry careers in general. The panel is followed by a keynote speaker who discusses research related to the love of chemistry theme. After lunch, the schools compete in a chemistry demonstration competition. The symposium ends with tours of the chemistry building and a mixer.

The club participates in several other professional and academic activities. The WSU ACS-SA attends ACS National Meetings and presents Successful Student Chapter Posters. The club attends the Detroit Local Section Student Member Meetings and ACS Regional Meetings when possible (*3*). Additionally, the WSU ACS-SA competes annually in the Battle of the Chemistry Clubs, a multi-school event held in January. Beyond participation in the events described above, the WSU ACS-SA chapter organizes its own outreach events such as Cub Scout Chemistry Day and high school chemistry tours. Members also volunteer for ACS Detroit Local Section outreach events. The ACS Detroit events include Earth Day demonstrations at the Detroit Zoo, Girl Scout Chemistry Day, and Zoo Boo (a Halloween-themed evening at the Detroit Zoo with costumes, candy, and hands-on chemistry demonstrations).

Campus events are wonderful opportunities for recruiting new members. WSU hosts three annual recruiting events for student organizations: Festifall for freshmen before fall classes begin, Student Organization Day for all students in September, and Winterfest for all students during the winter semester. The WSU ACS-SA participates in all three of these events and performs a liquid nitrogen ice cream demonstration during Festifall and Student Organization Day. These two events recruit the majority of new members each year, and the liquid nitrogen ice cream is a crowd favorite. Our members are trained to safely handle liquid nitrogen, and we make instant ice cream for the crowd with enough ice cream for everyone to sample. It is fun for the chapter and the audience. The chapter's members dressed in lab coats and safety glasses, free ice cream, and the spectacle of the liquid nitrogen demonstrations enable our club to stand out from the other clubs when competing for the attention of potential new members. Flyers with meeting details and scheduled events are handed out with the ice cream, and potential new members have a moment to hear about the club as they eat their dessert.

Some of the events that the WSU ACS-SA organizes are just for fun. In 2015, members used chemistry to make their own holiday ornaments to hang on a tree in the Chemistry Department. The club has toured Behind-the-Scenes at the Detroit Institute of Arts to learn about the analytical chemistry involved in art restoration and authentication. Annually, during the winter holiday break, the WSU ACS-SA holds its holiday social where students ice skate together before enjoying dinner at a restaurant in downtown Detroit. In addition, at the end of the school year, the WSU ACS-SA has established a tradition of a dry ice curling competition for the entire Chemistry Department. Undergraduates, graduate students, faculty, and staff vie for the department championship by testing their skill in the atrium of the chemistry building. Finally, the club occasionally organizes smaller events and often volunteers to help with events hosted by the College of Liberal Arts and Sciences, Chemistry Department, and the university.

Raising Funds

As events are planned, the question of funding generally arises (Figure 2). While some activities are free to plan, several of the events require a substantial influx of money to operate. The bulk of the chapter's funding is from three different sources. One source of grants is the ACS undergraduate office (*4*). The ACS grants available to undergraduate chapters are listed on the ACS Student Chapters section of their website. The WSU ACS-SA has received two types of ACS grants: National Meeting Travel Grants ($300) (*5*) for travel to the ACS National Meetings and Inter-Chapter Relations Grants ($750) (*5*) that have supported the For the Love of Chemistry Symposium. The other current ACS undergraduate grants include ACS Starter Grant for New Chapters, the Community Interaction Grant, the New Activities Grant, and the Undergraduate Programming at Regional Meetings Grant. The second major source of funding for the WSU ACS-SA is the Dean of Students Office at WSU that offers grants for student organizations at WSU through the Student Activities Funding Board. That board has approved annual grants for the Faculty–Student Mixer and the For the Love of Chemistry Symposium. The third and largest contributor to the club is the WSU Chemistry Department. Much of this support has been generously approved by Professor James Rigby, the chair of the department, who is a strong advocate for the club. The department allows the club to use its facilities for events and donates money to support the club's programs. The department financially supports the For the Love of Chemistry Symposium, the Faculty–Student Mixer, and trips to the ACS National Meetings. In addition to the three major sources of funding, the WSU chapter has received money through smaller fund-raising efforts including membership dues, small donations, bake sales, cleaning labs, and T-shirt sales. For more ideas, we recommend talking to other chapters about their successful ideas, consulting the ACS website, and reading *inChemistry* magazine for suggestions of innovative fund-raising approaches.

Figure 2. Flow chart for the steps suggested to setup meetings, plan and participate in events, and find funding.

Recruiting Members

At the beginning of the third year, the WSU ACS-SA spent all summer planning events and ways to fund them, but when the fall semester started, the main goal was to recruit new members. One of the most important elements for the chapter's success is recruiting and maintaining student members. The WSU chapter strives to maintain a diverse club, so membership in the chapter must be attractive to students of all undergraduate levels from new freshman to graduating seniors. From a club administration point of view, younger students might turn into multi-year members and future multi-term board members. Older students often have experience to share with other members as potential board members. Moreover, from the club's perspective, a membership that is representative of the entire chemistry undergraduate student population is ideal. At WSU, the club hosts a variety of activities, some of which may appeal to different types of students or undergraduate levels. The WSU ACS-SA chapter helps students who are interested in meeting people who can provide advice regarding classes and majors as well as making like-minded friends and potential study buddies. Most members appreciate the chances to learn and show leadership, to acquire professional skills, and to volunteer and resume-build.

In addition to providing social and professional opportunities, membership must be affordable for the students, and there must be a value to paying the chapter dues as well as the ACS National membership dues. The chapter membership for the WSU ACS-SA is currently $10. Some benefits of the WSU Student Chapter are for members only. For example, while on-campus events are open to all students, many off-campus events are only open to members.

For WSU students attending ACS National meetings with the chapter as paid chapter and national members, the WSU ACS-SA has been able to pay all hotel expenses. This financial support allows students to attend meetings and present their research or just come to learn more about careers in chemistry and visit a new city. The club also currently organizes annual tours of chemical and pharmaceutical manufacturing facilities, and these tours are only available to paid members. Furthermore, only paid members have voting rights for elections or meeting decisions, and they are afforded the first priority to attend events with a limited capacity. Finally, the WSU ACS-SA requires that any student that wishes to run for an executive board position is a paid chapter and national member prior to the election.

Communication and Promotion

Membership recruitment at WSU is closely linked to club communication. The Secretary is responsible for emailing all members and other interested WSU students to promote events, membership, and other club news. At every meeting and event, the email addresses of all the students present are collected. A comprehensive email list of interested students and members is an invaluable asset to any ACS chapter. The WSU ACS-SA has recognized that communication is essential to the chapter's success. Regular email messages inform students of upcoming meetings, events, internships, and volunteer opportunities. The WSU ACS-SA chapter endeavors to send minutes from the most recent past meeting promptly to update absent students of upcoming activities. Student attendance at meetings varies due to conflicting class schedules, work, exams, and other issues, so communication allows all students to participate as much as they choose. The club uses the email list judiciously and limits the club emails to those that are directly useful to club members because too many emails can cause students to request removal from the email list. Additionally, the WSU ACS-SA chapter was able to obtain help from the department to use departmental email lists to send relevant emails to all Chemistry students, faculty, graduate students, and staff. The club uses this option to invite the entire department to participate in the annual end of the year dry ice curling departmental competition. For invitations to faculty, we found that in-person invitations deliver a higher turnout than email invitations. For example, we organize the Faculty–Student Mixer that helps chemistry majors learn about undergraduate research. For the Faculty–Student Mixer, members of the WSU ACS-SA deliver printed invitations by hand to each faculty member after sending email invitations. The faculty commitment to attend the mixer increases substantially when students ask them personally to attend.

Email is only one way to circulate information about the club. Spreading the word about the latest events of the WSU ACS-SA is a multimedia effort. Flyers, emails, and visits to chemistry classes yield good results. Digital media tends to provide no-cost and low-effort options to boost awareness about the club's activities, and the WSU club largely focuses on its website and Facebook page. The WSU ACS-SA attempts to take pictures of all the activities each year. These pictures are perfect for the website, Facebook, the annual report and future promotions. For the club's website, a free Google site for the webpage combined with a static page on the WSU Chemistry Department's website are used, but there are many free ways to create websites. Facebook also works well for the WSU club at no cost. When a WSU member dedicates time to regular Facebook postings, attendance at meetings and events rises. The WSU ACS-SA chapter has collaborated with the *South End* (the school paper), the university event calendar, and the public relations offices in the Chemistry Department, College of Liberal Arts and Sciences, and the university itself. Furthermore, all of the public relations offices at WSU have active Facebook pages that promote chemistry club events and achievements.

Finding a Physical Home Base

During the chapter's third year, membership increased to 70 paid members, new grants were approved, and an ambitious series of events were planned for the year. Since May 2013, the new board had focused on collecting supplies for the chapter. Prior to the third year, the chapter had relied on the board members to purchase items at their own expense when needed. However, nothing was left with the club, so the club had no possessions of its own. Small items that were donated to the club were often lost when board members graduated. The Chemistry Department took notice of the expansion of the club and offered an office in the Chemistry Building. A physical office provided a place to accumulate assets that belonged specifically to the club. All of the equipment, demonstration supplies, decorations, flyers, membership applications, and office supplies of the club are now secured and easily available. Once an office was acquired, the club began to accrue supplies and save money because small items were lost less often, avoiding unnecessary replacements. The office was more than just storage; it also established a home base and club presence in the Chemistry Department. Finding a permanent physical home was necessary for the development of the WSU ACS-SA, and the Chemistry Department is definitely home for the club.

Designating Event and Committee Chairs

Once the chapter began to increase the number of annual activities it organized, a significant contributor to the chapter's success came from assigning responsibility to individual officers and members for all events and important undertakings. Early in the re-established club, certain details were forgotten or overlooked, or students assumed that someone else would take responsibility for a particular issue. These lapses in attention to detail meant that the club often ended up scrambling at the last minute. Since our third year, for each major event, an event chair or event co-chairs have been assigned to oversee the running of the event. Each event is divided into committees for the most important components, and each committee has its own chair, such as a decorations chair, food chair, correspondence chair, or public relations chair. The method of assigning personal responsibility for all the details of an event to individual members has been an organizational strength for events hosted by the WSU ACS-SA. The chair positions are usually designated based on experience and member requests. For the larger events, such as the Faculty–Student Mixer and the For the Love of

Chemistry Symposium, co-chair teams are usually chosen to pair an experienced board member with a newer member. This allows the newer member to gain experience for the next year and ensures that event knowledge is retained within the chapter. Additionally, chair titles enable members to demonstrate their leadership abilities on their resumes or during interviews because the titles suggest leadership over an important event or committee.

Strengthening the Relationship with the Chemistry Department

As the club grew, it actively tried to connect with all aspects of the Chemistry Department to become an integral part of the department's identity. ACS members volunteered to help at Chemistry Department events. ACS Chapter events occured in the department on a regular basis allowing everyone to become familiar with the chapter. The chapter reached out to all faculty members to include them in the Faculty–Student Mixer and other events. Faculty members, graduate students, and chemistry staff support the club's bake sales, mixers, symposia, and everyday events. Members of the WSU ACS-SA began visiting chemistry classes to promote their club and events with the permission of the faculty. Overall, connecting with the entire Chemistry Department has given strength and support to the developing club while providing an eager group of volunteers to help at Chemistry Department events.

Volunteering on Campus

Beyond the Chemistry Department, volunteering to help around campus has forged bonds and relationships between WSU ACS-SA members and different parts of the university. Our liquid nitrogen ice cream demonstration has been requested to be included as part of the Chemistry Department Graduate Research Symposium, a College of Liberal Arts and Sciences Symposium for potential donors, and Wayne State University's Open House Days for high school and transfer students. Student members have also assisted the college with other events including campus tours and informational sessions for potential undergraduate researchers.

Connecting with the Local ACS Section

One of the best strategies to help a new chapter is to contact the local ACS section (Figure 3). The ACS Detroit Local Section has offered the WSU ACS-SA chapter invaluable information on speaker choices, contact information for other ACS student chapters in the area, fun ACS events and talks, summer internship possibilities, as well as volunteering and networking opportunities. For WSU and other local student chapters, the Detroit Local Section hosts fun monthly lectures at their Younger Chemists Committee Brewing Chemistry Nights; organizes Kids and Chemistry events such as Zoo Boo, Earth Day at the Zoo, and Girl Scout Chemistry Day; donates funding for a Local Student Section Meeting in the Spring; and hosts various other lectures and social events. Furthermore, the ACS Local Section can connect students with industry representatives that can help arrange educational tours of companies in the chemical industry or academic facilities.

Figure 3. Flow chart to the steps suggested for a chapter to recruit members, develop and grow inventory, and establish a reputation.

Connecting with Other ACS Student Chapters

The ACS Local Section is a wonderful resource for meeting other ACS student chapters, and the WSU ACS-SA has benefited greatly from interacting with other student chapters. When the chapter attends an ACS National Meeting, members already know students from other schools and have had big group dinners. The WSU chapter loves to mingle with other chapters, as exemplified by the third annual For the Love of Chemistry Symposium being attended by seven schools in 2016. For the WSU chapter, interchapter events lead to future professional networking partners, current friendships, friendly rivalries, and a lot of fun. The WSU ACS-SA attempts to meet as many other chapters as possible including those in the local section area and many much farther away.

Building on Success and Generating Continuity

A chapter that is running well and wants to continue after active members graduate must plan for the future and leadership succession. Professor Mary Kay Pflum remarked that in the past, the club had formed, ran well for a period of time, and then disappeared more than once during her years at WSU prior to the new version started in 2011. Professor Pflum praised the students involved in the previous chapter as ambitious and hard working. So what happened? Why had the club failed? The club seemed to rely on a handful of motivated individuals, and when those individuals graduated, the club faded. So how does a new club prevent this fading from happening again?

The WSU ACS-SA has made the future of the club a priority. Board members actively recruit and train their replacements. A digital Dropbox account was created using the club email address, rather than personal email accounts, to establish a club document archive that would last beyond graduations. Board members are encouraged to store copies of all club-related documents, grants, posters, and flyers in this centralized location so that future boards will not have to reinvent the club from scratch every year. When recruiting new members, the board strives to have a density of younger members that can run for the board for more than one year in the future, providing some generational memory. It is a core value of the WSU chapter that new members need to be recruited and urged to run for board positions. To have a chapter that lasts, the board plans for the future every year. Elections were placed in February so that incoming board members would have at least two months of training from their outgoing counterparts. Additionally, the chapter holds a transition meeting at the end of the winter semester with the outgoing and incoming board members. The transition meeting facilitates a direct conversation between the two boards regarding an evaluation of performance from the previous year, an itemization of any leftover tasks, an acknowledgement of any commitments already made for the next year, and a discussion of the potential events for the next year. This meeting ensures that the chapter is functioning before school begins in the fall and can repeat previous successful events and try new things.

Another way to build for the future is to periodically review and revise the official governing documents. With a chance to reflect on the chapter's experiences, these documents improve over time. For example, the WSU ACS-SA's constitution and bylaws were revised to include a code of conduct, explicit election regulations, conflict resolution suggestions, and methods to remove and replace officers. Each of the additions arose from situations where there was no clear path to resolve issues. Having unambiguous regulations provides a structure and guidance for how the chapter needs to proceed. Now, all new board members are required to read and sign a copy of the constitution before their term, so that they are aware of the expectations of their position. A beginning chapter might choose to include these types of provisions in their initial constitutional draft.

Further constitutional revisions have changed the number of board members from the initial five to nine, then eleven, and then seven board positions while changing, inventing, and adapting the roles for each executive. The current seven positions are President, Vice President, Secretary, Treasurer, Assistant Treasurer, Public Relations Coordinator, and Outreach Coordinator. The chapter had thought that a larger board would allow for more events and a more active chapter. However, the larger board led to a higher turnover rate with more resignations and removals. The most recent revision to seven board members has worked to reduce turnover and focus on a core group of dedicated members to compose the executive committee. The WSU chapter has recognized that change is important, and that evaluating what is working for the chapter, and what is not, allows the chapter to correct its course to ensure that its goals can be met.

Another method for the chapter to build on past success is to improve the chapter every year using what the group has learned (Figure 4). When an ACS Student Chapter report is reviewed at the end of the year, usually the reviewers provide concrete examples of things to do to improve the chapter's evaluations in subsequent years. These suggestions are from reviewers who have had experience with many chapters and have the best intentions in mind for the chapter's success. Every year, the club has attempted to respond to the suggestions from the previous annual reports, and this effort has improved the chapter. One of the best examples from the WSU ACS-SA experience occurred after the review of the ACS student chapter report from the third year. The third year had made a large improvement in most areas, but the reviewers felt that more outreach activities needed to be held with children from kindergarten through twelfth grade. The club volunteers each year to help with the ACS Local Detroit Section Girl Scout Chemistry Day, but a second event was added in 2014, a Cub Scout Chemistry Night. The new event was designed, organized, and managed by the club. The addition of this event contributed to a better review for the chapter report the following year.

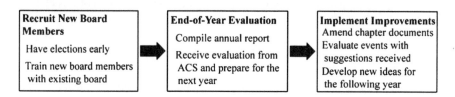

Figure 4. Flow chart to the steps suggested for chapters to build on the success from the previous year.

For the first two years of its new existence, the WSU ACS-SA received the ACS Honorable Mention Student Chapter Award. In the third year, after a massive improvement effort by the board, WSU earned the ACS Commendable Student Chapter Award. The WSU chapter was thrilled to receive the ACS Outstanding Student Chapter Award for its fourth year after reactivation. The improvement in the evaluations was connected to the chapter's direct effort to add new events while perfecting established events to create a more well-rounded chapter. Adding new activities every year grows the chapter into a better student organization while providing new interesting events for multi-year members. In 2016, WSU's fifth year since being re-established, several new events are planned.

The chapter is continuing to develop in its fifth year. How can all those who have participated in the chapter and care about the organization help safeguard the club's future? The WSU chapter has formed an ongoing collaboration between the current board members and graduated board members to foster continuity. Several alumni are available to help the current board by answering questions, sharing documents, and providing advice and support. Some of these alumni have even joined the ACS Detroit Local Section and help facilitate communication between the two organizations. ACS members, both current and graduated, care about the future of their club and are willing to devote time to help it become a permanent ongoing part of the WSU campus for years to come.

A final idea that the club is working on to generate continuity is to create an alumni network of WSU ACS-SA graduates. An alumni network could provide networking opportunities, career advice, and possibly future internship and job opportunities. In a few years, a new event could take place on an annual basis at the WSU Chemistry Department that would be a Student–Alumni Mixer. Once graduates have had time to prosper in the job market, a list of dedicated alumni could help the chapter continue to grow and flourish.

Conclusion

The WSU ACS-SA has blossomed over the last five years into a club that is active on campus, with the ACS Detroit Local Section, and with other ACS Student Chapters. The students in the club have diligently worked to develop and expand the club. From its small, enthusiastic beginning, to its more established present, the club has been shaped by student ideas and goals. The path to a stronger and better chapter has been a learning process for the club leading to constant self-evaluation followed by plans for improvement. The ability to build on success, learn from mistakes, and prepare for the future as a group has been a strength of this club.

References

1. The ACS designated undergraduates as "student members" in the late 2000s; however, when the WSU club re-activated, the moniker "student affiliates" was used. Consequently, throughout this chapter, we chose to use the name of the club that is in our constitution and our official designation for both WSU and the ACS.
2. ACS Start/Reactivate a Student Chapter. https://www.acs.org/content/acs/en/education/students/college/studentaffiliates/start.html (accessed July 5, 2016).
3. Detroit Local Section of the ACS. http://detroit.sites.acs.org/ (accessed July 5, 2016).
4. ACS Student Chapter Grants. https://www.acs.org/content/acs/en/funding-and-awards/grants/acscommunity/studentaffiliatechaptergrants.html (accessed July 5, 2016).
5. Values are from 2012–2016.

Chapter 2

Key Components of a Successful and Sustainable Student Chapter at a Public-Regional University

Steven R. Emory* and Elizabeth A. Raymond*

Department of Chemistry, Western Washington University, Bellingham, Washington 98225
*E-mail: steven.emory@wwu.edu; elizabeth.raymond@wwu.edu

We describe the ways in which Western Washington University's (WWU) Student Chapter of the American Chemical Society (SC) has established and sustained a dynamic and inclusive environment that is responsive to and meets the needs of its student members. The SC annually organizes and hosts a significant number of social (e.g. "Chempalooza", "Costume Bowling", and "Chemistry Trivia Night"), professional development (e.g. "Who, What, When, Where, How, and Why of Graduate School", "Undergraduate Research Opportunities", and "College to Career Discussion Panel"), and outreach (e.g. "Wizards @ Western", "Compass-2-Campus", and "Chemplosion!") events. These activities help build a cohesive SC while addressing the needs of its members. Combining clear organizational structure, a strong and supported membership base, and a focused and needs-aligned program enables the SC to be both successful and sustainable.

Introduction

Student-oriented science organizations, such as the Student Chapters of the American Chemical Society (SC), offer an additional experiential platform beyond traditional lecture, laboratory, and research to further student engagement and learning in the sciences. A successful student chapter can have a tremendous impact on both a department and an institution as a whole. This includes, but is not limited to, increased student retention and further engagement of students in scholarly activity (*1, 2*). A question that arises is what truly defines a successful chapter? As student chapter co-advisors at Western Washington University (WWU) we have struggled with this question and have come to answer it as follows: a successful SC meets and responds to the needs of its student members. However, this answer is somewhat inadequate. A successful SC is also sustainable over time, which is a much higher standard. A chapter needs to help invigorate its members, not burn them out. This is a challenge many programs face and it requires a careful balance of SC activity and resources. At WWU, we have created a successful *and* sustainable chapter by addressing three key support components (Figure 1). First, we have established a clear leadership structure that enables better chapter organization, which helps sustain and support chapter activity throughout the entire year. Second, we have created an inclusive environment that promotes strong support from students, faculty, and administrators. Third, we have built a program that addresses the needs of our student members at various stages (e.g. 1st-year students vs. seniors) during their time at WWU. We continually evaluate and refine aspects of the program based upon member feedback and participation.

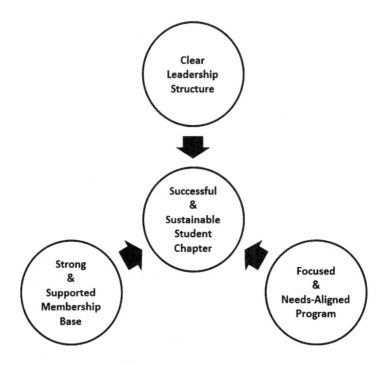

Figure 1. Key Components of a Successful and Sustainable Student Chapter.

Clear Leadership Structure

Leadership structure is an essential component for ensuring the sustainability of a successful SC. There are many types of structures that can work at a given institution (*3*). At WWU, we have found that a SC executive committee (Table I) that works in collaboration with the faculty chapter advisors is highly effective. Each executive committee member has clearly defined roles and responsibilities. Table I shows the current delegation of these responsibilities; however, duties may change based on the make-up of the specific SC executive committee. Assessing leadership abilities is an important task performed by the SC faculty advisors. This informal process involves discussing individual strengths, weaknesses, and opportunities for professional growth with each student leader at the beginning of the academic year. *Weekly* executive committee meetings are held to discuss chapter issues and to plan the agenda for the upcoming *weekly* general membership meeting. While these executive meetings are often short (~ 30 minutes), they are critical to how the chapter functions. Chapter issues (e.g. upcoming events, chapter goals, etc.) are discussed in detail at these meetings. This prepares the student leaders for the general membership meetings, which run more efficiently as most of the small details have already been discussed. Short, focused, and productive membership meetings are important to maintaining broader student interest and participation in the SC. Students feel that their time is respected and that their input is valued. These short and regular membership meetings really help the chapter stay on task and maintain positive momentum throughout the year.

President vs. Co-Presidents Leadership Models

The WWU SC has been successfully led by either a president or co-presidents. However, as the SC has grown, the responsibilities of the president have needed to be borne by more than one person. Rather than have a very strong vice-president, the chapter has opted to elect two co-presidents. Each spring, the chapter discusses and votes for candidates for the executive committee. Decisions on candidates are usually reached by consensus. For continuity, it is preferable to have one of the presidents to be of junior or sophomore academic standing so that the possibility of a co-president serving two consecutive terms exists. For this joint-leader structure to function well, each co-president must have very clear responsibilities. For example, one is assigned to lead the weekly membership meetings and the other is assigned to coordinate recruiting efforts. We have found that this dual leadership structure promotes cooperation and collaboration, while distributing the responsibilities and workload. In addition, this model fits the collaborative culture of our department and institution.

Table I. Student Chapter (SC) Executive Committee.

Position	Responsibilities
Co-Presidents	• Lead chapter meetings. • Represent the chapter at events. • Coordinate recruitment efforts. • Monitor chapter climate. • Encourage participation and cooperation. • File chapter report with ACS.
Treasurer	• Maintain a current budget for the chapter. • Involved in chapter fundraising.
Secretary	• Record minutes of meetings. • Maintain chapter roster. • Maintain chapter activities calendar.
Outreach Coordinator	• Maintain a list of chapter volunteers. • Communicate with and encourage volunteers. • Disseminate information about chapter events (e.g. e-mail, flyers, and social media).
Faculty Advisor(s)	• Oversee chapter governance. • Provide advice on policy matters. • Help students develop organizational and communication skills. • Provide continuity for the chapter. • Facilitate communication between the chapter and department/university units.

Chapter Organization and Communication

Maintaining organization requires clear record keeping and some level of redundancy. For example, we use a laboratory notebook called "The Red Book" to record chapter meeting minutes. These include the members in attendance, agenda items discussed, dates of events, as well as decisions made by the chapter. We also maintain a large dry-erase calendar in the student library where chapter meetings occur. The calendar is updated during each meeting as the chapter members make decisions. In addition, the faculty advisors maintain their own notes and calendars as a back-up. This information then gets transformed into advertising flyers as well as announcements on social media. Ultimately, all of these sources are used to write the end-of-the-year chapter report that is evaluated by the ACS (4).

The SC uses a variety of methods to communicate with its members, including weekly e-mails containing the agenda of the upcoming membership meeting, flyers in the chemistry building advertising events, and posts on the chapter's Facebook page (www.facebook.com/wwuchem). The Facebook page is useful for advertising events, keeping in touch with alumni and friends of the department, sharing interesting chemistry-related news stories, and also serves as a repository for event photos and information. We have found that the most effective way to publicize events is for members to personally invite their friends and classmates; the active in-person invitation (e.g. "I am going, are you?") is much more compelling and effective than the passive viewing of event e-mails or flyers.

Strong and Supported Membership Base

Communication, Recruitment, and Retention

Effective communication is an essential component of a successful and sustainable SC. In-person interactions are still the most important aspect of any SC, thus the chapter holds weekly membership meetings. Meetings are held at the same day and time every week throughout the year, every year. We used to poll members quarterly in an attempt to find an optimal meeting time. However, this process was cumbersome and sometimes confusing. We have found that setting a regular time, which does not change from year to year, makes the chapter meetings predictable and more accessible to students and their ever-changing schedules. Our regular meeting day, Tuesday, is early enough in the week to facilitate planning events. Our regular meeting time, 5:30 pm, is after most chemistry labs have ended, but early enough that students do not have to come back to campus to attend. All students are invited and welcome to attend membership meetings. While we strongly encourage students to become members of the ACS, this is only required for executive committee members. We also do not charge a membership fee for participation in the SC. As a public-regional university, many of our students struggle financially, so any membership fee can severely limit participation. Finally, we are a very inclusive chapter that regularly recruits and welcomes non-chemistry majors as full members. In addition to the small number of declared chemistry and biochemistry majors, our courses contain a large number of students who have either not yet declared a major or have declared a non-chemistry major. We are able to attract these students as members because of the organization, activities, and inclusive environment we create. Specific efforts to recruit non-majors include: 1) providing faculty teaching courses with PowerPoint slides to advertize SC events at the beginning of their classes, 2) collaborating with other STEM departments and clubs on events, and 3) encouraging members to personally invite fellow students to social, professional development, and outreach events. This broad recruitment approach is very important to the success of the SC. The key for *retaining* majors and non-majors alike is engaging them in activities they find meaningful. The resulting diversity of our membership has further facilitated collaboration with other departments and student-led science clubs.

We are deliberate in inviting new students to join the chapter. We hold a high-visibility kick-off event in the fall, Chempalooza, where we make liquid nitrogen ice cream and celebrate the accomplishments of the previous year. We provide students with information about the SC and the ACS. We make it very clear that everyone is welcome and wanted. Within one week, we follow this kick-off event with a regular membership meeting at which students discuss the goals and plans for the upcoming year. In order to retain members, they must feel engaged and valued. This is an iterative process, so we have developed an annual program that follows the cycle outlined in Figure 2.

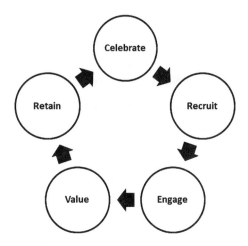

Figure 2. Recruitment and Retention Cycle. Recruitment and retention is an iterative process. Initial recruits must be engaged in activities that they perceive have value, which improves retention. In turn, the SC uses social activities to celebrate these successes and recruit more members.

Social Activities

Rather than hosting a large number of social events, our chapter has been much more successful organizing a few large (~ 100 people) events that have become annual SC traditions (Table II). These events are free to everyone. We encourage all students to participate and we promote faculty-staff-student interactions in an informal setting. Having annual traditions creates structure for the SC and helps establish a chapter identity. Recurring events can be refined annually and are easier to organize. The chapter is able to build upon success and grow these events every year. In addition, a sense of chapter history or even legend is established through the retelling of stories from previous years' events. This is not a static program; the chapter works to refine events every year and introduces new events when appropriate.

Fundraising and Branding

In order to host our social, professional development, and outreach events during the year, the chapter needs a considerable amount of funding. We are quite deliberate in our fundraising efforts, choosing to raise money in a small number of profitable ventures, rather than constantly asking our members for both their time and money. We have determined the profit-to-time ratio of bake sales and other small fundraisers is too small for our students' busy schedules.

Table II. Chapter Social Activities.

Quarter	Activity	Description
Fall	Chempalooza	Annual kick-off event to welcome new and returning students. Celebrates past accomplishments and recruits new members.
	Costume Bowling	Students select a costume theme (e.g. Science vs. Science Fiction, Librarians vs. Barbarians, etc.) for this popular team bowling competition.
	Mix It Up	Collaborative event hosted by science clubs at WWU. Event is designed to promote inclusivity in the sciences.
Winter	Chemistry Trivia Night	Popular team competition. In addition, members vote to select a new SC t-shirt design for the year.
	Chemistry Tea	Students, faculty, and staff gather and socialize at this informal event.
Spring	Department Picnic	Students, faculty, staff, and friends of the department gather at this large, off-campus, day-long event. Food, fun, and games help celebrate the coming end of the academic year.
	Department Awards	Students are recognized for their academic and service achievements.
	Graduation Reception	Junior-level chapter members organize and host a post-graduation reception for graduating students and their families.
Summer	Lake Ann Pilgrimage	Members that are on-campus during the summer hike to Lake Ann in the nearby North Cascades National Park.

Our primary fundraiser is selling student-designed shirts. Due to the size of our undergraduate program, we typically sell about 150 shirts, depending on the popularity of the design(s). Each winter quarter we host a competition to select a new t-shirt design (Figure 3). Over the course of a few weeks, students submit draft designs to the SC faculty advisors. The faculty advisors ensure the submitted designs are appropriate before they are displayed for consideration. Attendees at the "Chemistry Trivia Night" social event vote for their favorite designs in between trivia segments. Sometimes several rounds of voting are required to reach a majority decision. T-shirt order forms and money are collected over several weeks. Collecting orders at specific times such as before department seminars or after SC meetings is effective. In addition to the new year's design, we also sell our classic WWU periodic table design and a selected past favorite.

Our second major fundraiser is the sale of customized beaker mugs with the WWU chemistry logo. While the upfront cost of purchasing the stock of mugs is large, the initial sales quickly recoup the cost and the overall profit margin is high. We typically order enough beaker mugs upfront to have enough on hand for several years.

Branding is an important aspect of our fundraising and communication efforts. T-shirt designs, beaker mugs, and communications all utilize a consistent and professional SC logo. The proliferation of shirts and mugs across campus also raises the visibility of the chapter and its activities.

Grants and Institutional Matching Funds

The chapter regularly applies for external funding (e.g. National ACS, Puget Sound Section of the ACS) to support its activities (5). The chapter has received several ACS National Meeting Travel Grants, ACS Community Interaction Grants, and ACS New Activities Grants over the past decade. Importantly, these external grants have helped to leverage additional internal matching funds from WWU. Securing internal matching funds also helps raise awareness of the chapter's impact at the department and university levels. The chapter has used these funds to support student travel to conferences, purchase and build demonstration equipment, and host external speakers. The process of preparing a grant proposal is a valuable experience for the student members involved. It teaches concise writing, budgeting, and planning skills.

Front   Back

Figure 3. Example of SC T-Shirt Design.

Focused and Needs-Aligned Program

A successful SC caters to the interests of its membership. However, to be sustainable requires the development of a focused and needs-aligned program. As the chapter has evolved, the members have expressed the need for a diverse set of professional development activities. We define professional development as an ongoing process by which individuals gain knowledge and develop skills for their career. Resources from the ACS such as the *ACS Guidelines for Chemistry in Two-Year College Programs (6)*, *ACS Guidelines and Evaluation Procedures for Bachelor's Degree Programs (7)*, and the ACS "College to Career" web resource (*8*) help inform and guide our professional development programming. Such opportunities supplement the existing undergraduate curriculum and can show students the value of a professional organization such as the ACS. Students involved in undergraduate research often receive some of this information through their research group network. However, students not involved in research often feel left out of such opportunities. Thus formal and informal professional development are important components of our annual SC programming. Over the course of several years, we have developed a comprehensive multi-year program that addresses many of the needs of our members.

Formal Professional Development Activities

These activities are intentionally structured over a two-year period to meet the needs of chapter members at appropriate stages in their undergraduate education. Table III and Table IV list and describe these activities and the time of year they are offered. The program is balanced with one or two professional development activities per academic quarter. In addition, the timing of these events is critical to ensure student engagement. For example, the "Statement of Purpose Workshop" (Table IV) is held in the fall when many students are working on their graduate school applications. The content and timing is what makes the programming relevant to the membership.

As a public-regional university, WWU has a significant population of students who transfer from community colleges. It is important to integrate these students into the WWU culture as soon as possible. The first activity of the year is our highly visible "Chempalooza" kick-off social event (Table II). The chapter serves cake and liquid nitrogen ice cream to all students (1st-year, transfer, newly declared majors, and returning students). This fun event celebrates our past year's accomplishments as well as welcoming and introducing new students to the chapter. At the event, we announce our upcoming chapter activities and make it clear that everyone is welcome and valued. Using social events like this to promote our professional development programming has been very helpful in recruiting and retaining members.

Year-1 Formal Professional Development Activities

Table III describes the first year of our two-year program. The first professional development event is the "Who, What, When, Where, How, & Why of Graduate School". This interactive forum presents information about graduate school, similar to the "Graduate School Reality Check" material distributed by the ACS (*9*). In addition, a panel of three faculty members and a current graduate student are present to promote discussion and answer student questions. It is our first professional development event so students receive information as soon as possible in order to consider educational and career options. This is time-sensitive information because many graduate school applications are due in the late fall or early winter.

Table III. Year-1 Formal Professional Development Activities.

Quarter	Activity	Description
Fall	*Who, What, When, Where, How, & Why of Graduate School*	Provides students with information about graduate school and the application process.
	ACS *Program-in-a-Box*	ACS webinar series that covers a variety of professional development topics.
Winter	*Panel Discussion: Undergraduate Research Opportunities*	Provides students with information about how to get involved in undergraduate research at WWU and outside institutions (e.g. NSF-REU, internships, etc.).
	ACS Program-in-a-Box	ACS webinar series that covers a variety of professional development topics.
Spring	*College to Career Discussion Panel*	Provides resources to students to explore career options. Alumni panelists share career experiences with current students.
	STEM into Graduate and Professional Schools	Another opportunity to learn about graduate/professional school programs. This is collaborative effort with other student-led science clubs at WWU.
Summer	*You, Me, & the GRE. Preparing for Subject Exam*	Summer program at which students take a practice subject GRE exam. Students attend faculty-led review sessions on exam subject areas.

In the late fall or early winter, the chapter hosts a panel discussion titled "Undergraduate Research Opportunities". Fourth-year undergraduate students involved in research comprise the panel. They present information about the impact of undergraduate research and how they got involved in these activities at WWU. In addition, panelists or moderators discuss how to apply for external research opportunities such as the National Science Foundation's Research Experience for Undergraduates (NSF-REU) program (*10*). Most undergraduate students are unaware of these opportunities in their early years. The late fall or early winter is a good time to offer this program as most faculty mentors and external programs accept applications for summer research positions in the late winter.

A recent event that the chapter has organized is the "College to Career Discussion Panel". This activity is based on the "College to Career" online resource offered by the ACS (*8*). The first part of the presentation (~ 15 minutes) demonstrates the diversity of resources (e.g. videos, career paths, interviews, etc.) available on the College to Career site. In particular, information about career prospects and education requirements for specific fields is explored. The remainder of the time is used for a panel discussion of career options. Panelists are recent (5-10 years) chapter alumni who are currently employed. We strive to assemble a diverse set of panelists who have taken a variety of career paths (e.g.

teacher, instrument technician, software developer, etc.). We also select panelists with different levels of education (B.S., M.S., or Ph.D.). Before the event, the chapter outlines a short set of questions they would like the panelists to answer. These initial questions are useful as they set the pace for the discussion and ensure that no single panelist dominates the conversation. Audience members are then able to ask questions directly of the panelists. This is turning into one of our most popular activities.

In the spring, we collaborate with other student-led science clubs at WWU to host the "STEM into Graduate and Professional Schools". This is similar to our fall presentation, but is broader in scope. Hosting it in the spring works well because it captures students that did not attend the fall presentation. This is especially important for transfer students as it can be challenging to get them involved initially.

The chapter also takes advantage of the ACS "Program-in-a-Box" webinar offerings (*11*). These webinars are offered several times throughout the year, and provide interesting content without much organizational cost. Beyond securing a room and advertising the event, the webinars require a very small pre-event time commitment. The varied topics of the webinars also often provide different expertise than is found on-campus at WWU. Student feedback indicates that webinars focused on topics typically not encountered in regular seminars or in classes, such as "Chemistry of Ice Cream", are the most popular.

The final offering of our Year 1 professional development program is called "You, Me, & the GRE". This endeavor is designed to help students prepare for the subject GRE exam, for which they often *incorrectly* believe they cannot study. Students start by taking a practice exam in June. After they receive itemized scores, a series of weekly subject area (analytical, biochemistry, inorganic, organic, and physical) review sessions are led by faculty members. Students also set up study groups to prepare for the fall GRE subject exam. A second practice exam is offered early in the fall, and we have observed some students increasing their scores by as much as 30 percentile points. Because the subject GRE has been moved earlier (September and October) in recent years, advertising this preparatory program raises student awareness of upcoming testing deadlines. This program has been very effective and it has strengthened students' graduate school applications; in recent years several have been awarded highly competitive NSF Graduate Research Fellowships (NSF-GRF).

Year-2 Formal Professional Development Activities

The activities offered in the second year (Table IV) of our two-year program build on the foundation of the first year. Students that missed the first-year events are still able to attend these activities. Again we start with the "Who, What, When, Where, How, & Why of Graduate School". Many students come to this session both years. Soon after this program, we offer a "Statement of Purpose Workshop". Students bring drafts of their statements and work together to improve their writing. We also provide a handout of tips and suggestions based on faculty experience and the ACS *Graduate School Reality Check* pamphlet (*9*).

Table IV. Year-2 Formal Professional Development Activities.

Quarter	Activity	Description
Fall	Who, What, When, Where, How, & Why of Graduate School	Provides students with information about graduate school and the application process.
	Statement of Purpose Workshop	Helps students write a statement of purpose for graduate/professional school applications.
	ACS Program-in-a-Box	ACS webinar series that covers a variety of professional development topics.
Winter	Professional Networking	Provides students with information and practice on how to effectively network.
	Making & Presenting a Poster	Provides information on how to create a high-quality poster and how to effectively communicate scientific information.
	ACS Program-in-a-Box	ACS webinar series that covers a variety of professional development topics.
Spring	College to Career Discussion Panel	Provides resources to students for exploring career paths. Alumni panelists share their career experiences with members.
	Resume & Job Search Workshop	Students learn to prepare a professional resume and refine interview skills.

In the winter, we collaborate with other student-led science clubs to offer a "Professional Networking" workshop that addresses the students' need to learn how to effectively communicate and connect with other professionals. We invite an expert from the College of Business and Economics to offer expertise. Students participate in activities designed to help them practice initiating professional conversations. This is a popular event that has helped develop student awareness of communication skills and has built bridges between students in different science clubs at WWU.

In the late winter, we offer a "Making and Presenting a Poster" workshop. This is timely because it occurs before the spring National ACS Meeting and before our ACS Puget Sound Section Undergraduate Research Symposium. In this workshop, students learn how to prepare a visually appealing poster and how to present scientific information in a clear and concise manner (*12*). One popular feature of this event is the discussion of "good" and "bad" aspects of example posters.

In the spring, we offer a "Resume and Job Search Workshop" that is specifically designed for students seeking employment after graduation. This is a relatively new activity for the chapter and is currently being refined based on membership feedback. We also provide students a sheet of online resources including the ACS *College to Career* web resource (*8*).

Community Outreach Activities

Outreach activities are a major passion for the chapter and hence are an integral component of our annual informal professional development activities (Table V). Students learn valuable skills from these public interactions including: public speaking, planning and organizing, teamwork, and leadership. Perhaps most importantly, they learn how to make science, chemistry in particular, relevant to the public (*13*, *14*).

Table V. Chapter Community Outreach Activities.

Audience Type	Activity	Description	Timing
Local Schools	Class Demonstrations[a,b]	Demonstrations selected to align with class's unit learning objective(s).	All Year
	Compass-2-Campus[a]	5th-grade students from underrepresented backgrounds visit WWU's campus. Demonstrations selected to engage students about scientific discovery.	Fall
	Science Fairs[b]	Chapter members visit elementary science fairs. Members interact with students and provide feedback.	Winter and Spring
Formal Demonstration Shows	*Wizards @ Western*[a,b]	Demonstrations selected based on a specific theme (e.g. Fire and Ice) and performed in conjunction with scientific explanations.	Winter
	Chemplosion![a]	Dramatic demonstrations (e.g. whoosh bottle, LN2 cloud, etc.) performed for a broad community audience.	Spring
Local Groups	(Girl Scouts, Boy Scouts, GEMS) Demonstration Shows[a,b]	Demonstrations selected to align with the group's current interests.	All Year
	Hands-on Activities[a,b]	Hands-on experiments selected to complement the group's interests and experience-level. All Year	All Year

[a] On-campus activity.   [b] Off-campus activity.

The chapter regularly interacts with local elementary schools throughout the year. Chapter members both visit schools and host science events on the WWU campus. The chapter works with local teachers to design science shows that reinforce concepts from class. For example, the chapter has acquired and refined a series of science demonstrations that focus on the states of matter and phase changes. This particular program can be tailored to any grade level (K-12), but has most recently been used with first and second grade classes. In addition, chapter members attend the science fairs of local elementary schools to interact directly with students. The chapter hosts large (> 100 people) annual events on campus designed for a more general community audience that include: "Chemplosion!" and "Wizards @ Western". The chapter also hosts community groups such as Girl Scouts, Boy Scouts, and GEMS (Girls Excelling in Math and Science) throughout the year. These groups perform hands-on experiments with chapter members in our laboratory facilities. These experiences are very rewarding as they are opportunities to serve the broader community. In addition, some chapter members have discovered a passion for teaching through these interactions.

In almost of all of our activities, we intentionally incorporate a social element allowing students to develop deeper relationships with each other and the participating faculty. This also provides informal time for reflection on the event and for brainstorming new ideas. The social component may be as simple as getting some ice cream or sharing coffee after an event. The strong interpersonal bonds forged through these interactions strengthen the chapter and improve member retention.

Assessment and Goal Setting

Annual assessment has become more important as our SC has grown. After every SC event, the chapter sets aside time at the next membership meeting to reflect on the activity. Ways to improve attendance, level of participant engagement, and the sustainability of the event are discussed. Detailed notes are recorded in the "The Red Book" and we review these notes before we host the event again. It is through this on-going assessment that the SC has been able to refine its activities and ensure they align with student needs. We have eliminated some activities (e.g. movie nights and bake-sales) because they were deemed an inefficient use of SC resources. Asking questions such as: "Is the activity worth the time and effort?" and "Does the SC have the resources (i.e. personal time) to invest?" are important when adding new activities and when deciding whether or not to continue or discontinue an event. During this process, chapter officers and faculty advisors must be cognizant of the potential for membership burn-out.

Having a clear and deliberate goal setting process has increased in importance as the chapter has matured. Each spring the SC faculty advisors meet with the current and future officers to reflect on the past year and to set goals for the upcoming year. This is typically done during the compilation of the annual ACS SC report (4). The chapter finds it useful to re-visit its goals throughout the year, because it reminds members of what the SC values and helps to keep the group focused. Having clear goals also provides a framework for allocating financial and time resources, which helps ensure the sustainability of the chapter.

Conclusions

Since its formation in 1969, WWU's Student Chapter of the ACS has continually grown and evolved along with WWU's changing student demographics. Through the process of self-assessment and refinement, we have been able to establish a successful and sustainable SC. Time and financial resource limitations are always present, but with a group of dedicated and engaged students and faculty, these challenges can be overcome. A clear leadership structure, strong and supported membership base, and a focused, needs-aligned program enable the chapter to be both successful and sustainable in a public-regional university setting.

References

1. Graham, M. J.; Frederick, J.; Byars-Winston, A.; Hunter, A.-B.; Handelsman, J. Increasing Persistence of College Students in STEM. *Science* **2013**, *341*, 1455–1456.
2. Wyatt, L. G. Nontraditional Student Engagement: Increasing Adult Student Success and Retention. *J. Contin. Higher Educ.* **2011**, *59*, 10–20.
3. Komives, S. R.; Lucas, N.; McMahon, T. R. *Exploring Leadership: For College Students Who Want to Make a Difference*; John Wiley & Sons: San Francisco, CA, 2007.
4. ACS Student Chapter Reports Home Page. American Chemical Society: Washington, DC, 2016. http://www.acs.org/content/acs/en/education/students/college/studentaffiliates/reports.html (accessed July 22, 2016).
5. ACS Student Chapter Grants Home Page. American Chemical Society: Washington, DC, 2016. http://www.acs.org/content/acs/en/funding-and-awards/grants/acscommunity/studentaffiliatechaptergrants.html (accessed March 31, 2016).
6. ACS Guidelines for Chemistry in Two-Year College Programs [Online]. Society Committee on Education, American Chemical Society: Washington, DC, 2015. http://www.acs.org/2YGuidelines (accessed July 20, 2016).
7. ACS Guidelines and Evaluation Procedures for Bachelor's Degree Programs [Online]. Committee on Professional Training, American Chemical Society: Washington, DC, 2015. http://www.acs.org/cpt (accessed July 20, 2016).
8. College to Career Home Page. American Chemical Society; Washington, DC, 2016. http://www.acs.org/content/acs/en/careers/college-to-career.html (accessed March 31, 2016).
9. Di Fabio, N., Ed. *Graduate School Reality Check*, 2nd ed. [Online]; American Chemical Society: Washington, DC, 2013. https://www.acs.org/content/dam/acsorg/education/students/graduate/gradschool/graduate-school-reality-check.pdf (accessed March 31, 2016).

10. Research Experiences for Undergraduates (REU) Home Page. National Science Foundation: Arlington, VA, 2016. https://www.nsf.gov/crssprgm/reu/ (accessed March 31, 2016).

11. ACS Program-in-a-Box Home Page. American Chemical Society: Washington, DC, 2016. http://www.acs.org/content/acs/en/acs-webinars/program-in-a-box.html (accessed March 31, 2016).

12. American Chemical Society. Surviving Your First ACS Undergraduate Poster Presentation. *inChemistry* **2012** (February/March), 16–18.

13. Bracher, P. J.; Gary, H. B. Chemists: Public Outreach Is an Essential Investment of Time, not a Waste of It. *Vision 2025: How to Succeed in the Global Chemistry Enterprise*; ACS Symposium Series; American Chemical Society: Washington, DC, 2014; Vol. 1157, pp 37–50.

14. Bursten, B. The Centrality of Chemistry: Our Challenges and Opportunities. *Chem. Eng. News* **2008**, *86* (1), 2–5.

Chapter 3

A Successful Student Chapter in the Energy and Medical Capital of the World

Elmer B. Ledesma* and Birgit Mellis

University of St. Thomas, Department of Chemistry and Physics, Houston,
Texas 77006, United States
*E-mail: ledesme@stthom.edu

The University of St. Thomas ACS chapter has been in
existence since the 1999-2000 academic year. Our chapter
provides many opportunities for the professional and personal
development of our students: volunteer activities serving the
university and general communities; professional networking
and career events; social events; and travel to national and
regional ACS meetings. The success of our chapter stems
from the commitment and dedication of officers, members, and
faculty advisors.

Introduction

ACS Student Chapter at the University of St. Thomas

The ACS Student Chapter at the University of St. Thomas, Houston, TX
(1), has been in existence since the 1999-2000 academic year. At the time it
was designated as a Student Affiliates Chapter. During the 2007-2008 academic
year, the designation changed to Student Chapter. The chapter is housed in the
Department of Chemistry and Physics at the University of St. Thomas. The
university is an independent, Catholic co-educational university, serving nearly
3500 students annually in credit and non-credit courses accredited by the Southern
Association of Colleges and Schools.

The University of St. Thomas is also a Hispanic-Serving Institution. Nearly
one third of undergraduate students and one quarter of graduate students are
Hispanic. Over one third of the student body receives some type of need-based

financial aid. Almost two thirds of our enrollment is female. All programs in the Science, Technology, Engineering and Mathematics (STEM) areas are undergraduate programs. Half of the entering freshmen indicate interest in STEM areas. The university has been expanding in the STEM areas in the last several years and we are currently in the process of constructing an integrated STEM and health sciences building.

Career Aspirations of Chapter Members

All of the members in our student chapter are majoring or have majored in STEM areas, *viz.* biology, bioinformatics, biochemistry, chemistry, and mathematics. The majority of our members (~ 90%) aspire to have a career in a health profession, whether in medicine, dentistry or pharmacy. This is not surprising considering that the world's largest medical complex, the Texas Medical Center, is less than 3 miles from campus and that since 2015, the University of St. Thomas is a member institution of the Texas Medical Center.

Though Houston is known for its world-class medical complex, it is far better known as the energy capital of the world, as all the major oil and gas companies have offices in and around the city. In contrast to those members with career goals in a health profession, there are very few members (< 10%) who desire, post graduation, to attend a graduate program in the natural sciences or engineering fields, or to obtain employment in the oil and gas or chemical process industries.

Chapter Membership

With the majority of our members having an interest in a health field, it might at first seem that the ACS Student Chapter at the University of St. Thomas would find it difficult to keep its members interested and to sustain its activities that primarily emphasize chemistry and chemical engineering concepts. Moreover, there are other campus clubs that are primarily focused on the health professions, and such clubs could potentially be attractive to our members, especially those who have an interest in such fields. Fortunately, our student chapter has not experienced a loss of interest by members nor a drop in membership.

Since our beginnings during the 1999-2000 academic year, we have seen an increase in membership from between 5-15 in 1999-2000 to between 25-35 in this current 2015-2016 academic year. In fact, we have had a constant 25-35 membership numbers for the past three academic years. Depending on their academic and other extra-curricular schedules, the amount of time that our members can devote to chapter activities may vary each year. However, such variability has not had a major impact on the success of our chapter.

A Successful Student Chapter

A chapter's success can easily be gauged by the chapter awards, if any, it receives from the ACS Society Committee of Education. Table 1 displays the awards that our chapter has received. As the table clearly shows, our chapter has received an award every year since the chapter started in 1999-2000. For the past

16 academic years, we received 6 Outstanding Awards, 5 Commendable Awards, and 5 Honorable Mention Awards. In addition, we have been twice recognized with a Green Chemistry Chapter Student Award. It is quite clear that the ACS Student Chapter at the University of St. Thomas is a successful student chapter.

Table 1. Chapter awards our student chapter has received.

Academic Year	ACS Society Committee of Education Award
1999-2000	Honorable Mention Award
2000-2001	Honorable Mention Award
2001-2002	Honorable Mention Award
2002-2003	Commendable Award
2003-2004	Honorable Mention Award
2004-2005	Commendable Award
2005-2006	Outstanding Award
2006-2007	Commendable Award
2007-2008	Outstanding Award
2008-2009	Outstanding Award
2009-2010	Outstanding Award and Green Chemistry Chapter Award
2010-2011	Commendable Award
2011-2012	Honorable Mention Award
2012-2013	Outstanding Award
2013-2014	Outstanding Award and Green Chemistry Chapter Award
2014-2015	Commendable Award

In addition to recognition by ACS, we have also been recognized by the university as an outstanding club on campus. In 2014, the university recognized the student chapter with 4 awards: (1) Outstanding Club Award; (2) Co-Programming Award for our work with the university's African American Student Union in screening the PBS documentary of Percy Julian; (3) The Marsha Wooldridge Citizenship Award to Sally Acebo, our 2013-2014 president; (4) Advisor of the Year Award to one of our faculty advisors, Elmer Ledesma.

To manage and maintain such a successful student chapter requires a good deal of commitment, organization, and communication. What follows in this book chapter are descriptions of the methods and events that the ACS Student Chapter at the University of St. Thomas has employed and conducted in order to ensure it to be a successful student chapter.

Chapter Leadership

Our student chapter has benefited from very good leadership throughout its existence. This leadership derives from faculty advisors and chapter officers.

Faculty Advisors

We are the current two faculty advisors for our chapter (see Figure 1). For the first 12 years since the chapter's beginnings, there was only one faculty advisor. Beginning in the 2011-2012 academic year, a second advisor was added due to our increased membership. Our advisors have been committed and dedicated to our chapter. Even though they are expected to excel in teaching and accomplish excellent research in their professions, our advisors have found the time to participate in chapter events. Although many events have been scheduled during the evening hours or during a weekend, our faculty advisors still found the time to participate and contribute.

Figure 1. Faculty advisors, Birgit Mellis (left) and Elmer Ledesma (right), at the national ACS meeting in Dallas, TX, March 2014. Photo courtesy of Jennifer Hoang.

Chapter Officers

We have had excellent chapter officers throughout our chapter's existence. The chapter officers form the core of the chapter. Having very good officers who are disciplined, organized, and dedicated to the chapter, is critical in order to ensure a successful student chapter. Our officer positions include the following: president, vice-president, treasurer, secretary, public relations, and demos coordinator. Figure 2 presents a photograph of our 2015-2016 officers.

Figure 2. 2015-2016 officers. Clockwise from top left: Alyssa Mullery (demos coordinator); Ana Hernandez (secretary); Pilar Zaibaq (vice-president); Danny Nguyen (public relations); and Angel Rivera (president).

Officers are recruited at the end of every spring semester. Nominations are requested every March at one of our chapter meetings. During this meeting, prospective candidates give a brief speech on why they should be voted. At the end of the meeting, members write names of candidates for the officer positions. The current officers and a faculty advisor tally the results and the new officers are presented at the last chapter meeting of the spring semester. To be eligible to be an officer, a student must at least be a sophomore and to be an active member.

Our officers are responsible for planning all events. This planning involves selecting a suitable date and time; selecting a location and if necessary to reserve a room; if food is going to be involved, what type of food to get; if the event is off campus, then organizing a car pool; and selecting what demonstrations to use. In addition to event planning, our officers, in conjunction with the faculty advisors, are also responsible for the planning of trips, especially to national ACS meetings. This trip planning mainly involves the selection of appropriate accommodation and transportation. Since much of the chapter activities are planned and run by the officers, it is very critical to find officers who are very well organized and committed to the chapter.

Chapter Communications

The communication of events to our members is accomplished through two ways: chapter meetings and Facebook. In the past we had a website. Due to costs in maintaining the website and the preponderance of social media, we discontinued the website during the 2011-2012 academic year and we now use Facebook for our digital communications.

Chapter Meetings

Chapter meetings are regularly scheduled every semester, with about 3-4 meetings per semester. The meetings are set before the start of every semester and are organized and planned by the officers. We attempt to schedule our meetings at times and days in which there are no other campus club meetings. Food is also provided, as our meetings are always scheduled during lunch. Figure 3 shows a typical scene at a chapter meeting.

Figure 3. Chapter meetings inform our members of past and future events. Photo courtesy of Jennifer Hoang.

The meetings are usually not that long (30 minutes maximum) with our chapter president informing members about past and future events. Sometimes we have had guest speakers talk about a particular topic of interest, such as preparation strategies for pre-health profession exams, graduate school as an alternative to medical school, or professors talking about their research projects.

Facebook

The wide usage of social media prompted us to make the switch in our digital communication from a website to Facebook (2). All members receive notifications about past and future events on Facebook in the form of simple text updates or as a graphic. Figure 4 illustrates an example of a graphic posted on Facebook that is used to notify our members about an upcoming meeting. The Facebook notifications reiterate what has been mentioned in chapter meetings. Photographs of past events are also posted on Facebook so members can stay informed about events they missed and have a vivid and fun reminder of ones where they participated. So far, we find Facebook to be quite useful in getting information out to our members quickly.

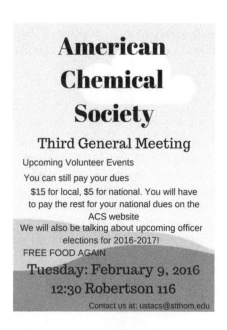

Figure 4. Facebook notification of a chapter meeting.

Chapter Events and Activities

Events are the heart piece in the life of our chapter. They help in forming a tight-knit chapter student community much better than chapter meetings. Through working and having fun together outside the classroom, participating members get to know each other well and make new friends. Events also let members earn volunteer hours as part of their eligibility (besides GPA) to wear Phi Lambda Upsilon (PLU) honor cords at graduation. PLU is the National Chemistry Honor Society (3). The university has had a PLU chapter since the 2011-2012 academic year. One of our faculty advisors, Elmer Ledesma, also serves as the faculty advisor for the university's PLU chapter. As another incentive, the three students with the highest amount of volunteer hours get honorably mentioned at the final chapter meeting at the end of the academic year and receive a small monetary prize.

Some events take place in the late afternoons or evenings after classes, but most are during the weekends. The big turnout for volunteering events during weekends (usually 10-15 students for the major events) shows how enthusiastic our members are about their chapter. Volunteer numbers for events are polled by the officers online. The event coordination lies predominantly with the chapter officers with assistance from the faculty advisors. At least one of the two faculty advisors participates in each event to support the students.

While some events are suggested by the faculty advisors, many great ideas come from the officers or the members. Every semester, we hold large events that involve many members and non-members, and smaller-scale events involving a few people. Our large events are typically ones we have conducted in the past and have become a tradition in our chapter and in our university. The smaller-scale activities are usually ones that we initially didn't plan for. What follows is a description of some of the large- and small-scale events that have been popular.

Lab Coat Tie-Dye and Barbeque

This is one of our popular events and is the first event we hold every academic year. Freshman and sophomore students who have recently purchased their clean, pristine and white lab coats have the opportunity to personalize it by tie-dyeing their coats. Figure 5 presents examples of lab coats that have been tie-dyed. This event is held on an afternoon beginning usually at 4:30 p.m. on a day during the first two weeks of classes when freshman students will not yet be requiring their lab coats. At the same time, we also hold a barbeque and members and non-members partake of the free food that is available.

Figure 5. Tie-dyed lab coats hanging to dry.

Freshman Club Fair

This event is organized by the university at the beginning of every fall semester. It is an event that gives freshman students the opportunity to find out about the various clubs and organizations present on campus. They can decide to join clubs and organizations that they are interested in. Our chapter always maintains a presence at this event. We have a table which we adorn with our banner, a sign-up sheet, a poster board describing what the chapter is all about, and numerous candy to distribute to students interested in our chapter (see Figure 6). All the officers and at least one faculty advisor are present at this event to answer queries and questions. During this event we usually have a large number of freshman joining our chapter.

Figure 6. Chapter officers, Danny Nguyen (left) and Angel Rivera (right), at the Fall 2015 freshman club fair event.

Education Rainbow Challenge

This event is partly organized by the ACS Greater Houston Local Section (*4*), and has been held every fall or spring for the past several years. Our chapter has always participated due to the connection we have with the local section: one of our faculty advisors, Elmer Ledesma, is a board member and is currently serving as an alternate councilor. This event is our busiest. It involves the participation of a large number of our members, who assist in giving chemistry and physics demonstrations to a very large number (~ 500-700) of middle-school students. The event is held on a Saturday morning between 8 a.m. and noon.

Due to the large numbers of students involved, other area organizations and companies also participate in the event. The main objective of the event is to spark an interest in STEM areas to middle-school students. Our chapter members who volunteer at this event are very enthusiastic and animated about describing the wonders of chemistry and physics. Figure 7 shows a photograph of two very dedicated chapter members at one of these events.

St. Augustine's Birthday

This event is organized by the university to celebrate the birthday of Augustine of Hippo, an early Christian theologian and philosopher. It is held every November on a weekend that either coincides or is closest to Augustine's birthday on November 13. We participate in this event with a table full of fun chemistry demonstrations (see Figure 8). It is held on Saturdays from around 10:30 a.m. until 2 p.m. when the activities wind down. Since this event is family-oriented, there are always lots of small kids who are very eager to see what the chemistry demonstrations are all about. In addition to small kids, our demonstration table is also visited by older children and parents, and on occasion, by a member of the university administration. St. Augustine's birthday is a very popular event and we usually have a good member turnout (between 10-15).

Figure 7. Kyana Garza (left) and Chinanu Chidi (right) at Education Rainbow Challenge (November 2014). Photo courtesy of Jennifer Hoang.

Figure 8. Chapter members preparing a chemistry demonstration at the St. Augustine's Birthday event (October 2015). Photo courtesy of Jennifer Hoang.

NanoDays at the Children's Museum

A main volunteering event for our chapter each spring semester is NanoDays at the Children's Museum of Houston. This weeklong event is sponsored by the Nanoscale Informal Science Education network (5) and takes place at hundreds of locations throughout the country. Its goal is to engage people of all ages in learning about nanoscale science and engineering and the potential impact on nanoscience on the future. Between 10-15 members participate each year. They volunteer for 3-4 hours either on a Thursday evening or on a Saturday and Sunday. After a short introduction by the museum personnel, students facilitate nano programming throughout the museum and assist with engaging young visitors in the various nano exhibits. Figure 9 shows an example of how our chapter members engage with the community at these events.

Figure 9. A chapter member demonstrates a nano exhibit to a young child. Photo courtesy of Jennifer Hoang.

Adopt-A-Beach

Our chapter co-sponsors this event every spring semester with Tri-Beta, the biology honor society on campus. Our members who volunteer for this event assist in the removal of garbage along an allotted section of a beach on Galveston Island. From this event, our members learn a great deal about sustainability issues, especially waste and its impact on the natural and built environments.

Earth Week

Every year during Earth Week (late April), we hold an event(s) to highlight environmental issues facing the world today. In the past we were primarily responsible for organizing and holding the event in collaboration with the ACS Greater Houston Local Section due to the relation we have with that local section through one of our faculty advisors, Elmer Ledesma. However, since it's founding in Fall 2014 we have also co-sponsored Earth Week events with the university's Sustainability Committee where one of our advisor, Elmer Ledesma, is a member. Last year we participated in two events: (a) co-sponsoring with the ACS Greater Houston Local Section a screening of "Disruption", a film that highlights the effects of climate change; and (b) co-sponsoring with the

Sustainability Committee a trash audit on campus, where members and a faculty advisor assisted in the sorting of the trash generated in a single day on campus into the categories of plastics, paper, metal, food, and trash. Figure 10 is a photograph showing the results of the trash audit that was conducted in April 2015.

Figure 10. Results from a trash audit event (April 2015).

Small-Scale Events

These events are ones that haven't been planned long term but are more spontaneous. One good example of these type of events are ACS webinars, which the ACS organizes. Depending on when the webinars are scheduled to take place, we usually hold these webinars either on the days after our chapter meetings or on an evening. For the most part, we get only small numbers of participants (< 12) for the two to three webinars we hold each academic year. Figure 11 shows a photograph of our chapter members attending the pheromones webinar that was held on Valentine's Day in February 2014.

Figure 11. Chapter members attending the ACS pheromones webinar on Valentine's Day 2014.

A small-scale event that was quite successful was a screening of the PBS documentary about Percy Julian. We co-hosted this event with the African American Student Union organization on campus in recognition of Black History Month in February 2014. Figure 12 shows the graphic that we posted on Facebook to let our members know of the event. We had a fairly decent turnout for the event with some 15-20 students. We also provided free pizza. Before screening the documentary, one of the faculty advisors, Elmer Ledesma, gave a brief presentation on the Civil Rights Movement in order to have some context.

Figure 12. Facebook notification graphic used for Black History month event (February 2014).

We also provide tours to area institutions such as Huntsman Coorporation, a chemical company with headquarters located in the Woodlands, TX, and the Texas Medical Center. The tours to Huntsman Coorporation are organized by one of the faculty advisors, Birgit Mellis, while the tours to the Texas Medical Center have been organized and planned by the chapter officers. These tours provide insight into the medical and non-medical careers that interest our members.

If a need arises, we sometimes hold workshops that are based on a general chemistry laboratory experiment that involve a formal lab report. General chemistry students are required to complete one formal lab report each semester. In the fall semester, the formal lab report is for the determination of the empirical formula of hydrous copper sulfate and for the spring semester it is for a kinetics experiment. These workshops are organized by the chapter officers and attendance is open to both members and non-members.

A faculty advisor is present to assist in the explanation of concepts. Figure 13 shows a scene where one of our faculty advisors, Elmer Ledesma, explain chemical kinetics during the kinetics workshop in the spring 2014 semester. These workshops are very well attended. We go over the basic chemical and physical concepts needed for the lab report and then we demonstrate calculation procedures using a sample experiment. It is hoped that students who attend these workshops clarify any queries that they may have and that they complement the knowledge they gain from lecture and lab classes.

Figure 13. Faculty advisor, Elmer Ledesma, explaining chemical kinetics at a kinetics workshop. Photo courtesy of Jennifer Hoang.

Undergraduate Research

In our department, it is a requirement that our majors engage in undergraduate research. In addition, they are also required to become national ACS members. The majors in our department are also chapter members. Our majors typically commence undergraduate research with the professor of their choice during their sophomore year. It is recommended that they stay with the same research project during the course of their undergraduate studies. As such, they can focus exclusively on a particular project that hopefully will lead to a journal publication.

Journal Publications

Since the beginning of our chapter in the 1999-2000 academic year, our student members are listed as co-authors in 16 peer-reviewed journal publications: 2 in *Industrial & Engineering Chemistry Research* (*6, 7*); 2 in *Energy & Fuels* (*8, 9*); 2 in *Journal of Molecular Structure* (*10, 11*); 1 in *Magnetic Resonance in Chemistry* (*12*); 2 in *Journal of Chemical Education* (*13, 14*); and 7 in *Journal of Undergraduate Chemistry Research* (*15–21*).

Conference Presentations

The vast majority of the research conducted by our members are presented at scientific meetings, especially the spring national ACS meetings. Figure 14 shows a photograph of two chapter members presenting their poster at a recent national ACS meeting. For every year since our chapter started, we have attended and presented at the spring meetings. We do not attend the fall national ACS meetings because they coincide with the start of our academic year. On occasion we attend regional ACS meetings. Over the last 16 years of our chapter's existence, we have only attended two ACS Southwest Regional Meetings.

Figure 14. Jennifer Hoang (left) and Alyssa Mullery (right) presenting their research at the national ACS meeting in San Diego, CA, March 2016.

Going to a spring national meeting involves some careful logistical planning on the part of both the officers and faculty advisors. If the meeting is out-of-state, we set a maximum number of about 12 students who can attend. We find this number to be reasonable in terms of management between the two faculty advisors and budget constraints. For in-state national meetings, we usually can take more students, as airplane costs, a major expense in addition to accommodation, are irrelevant since we drive to the location using the two university vans that are available. However, we try to limit student numbers to no more than 20. Our members find these national meetings very beneficial, since they often provide an experience in presenting research at a large, professional setting. Moreover, the students also enjoy visiting the sites a location offers and tasting the local cuisine.

Funding Sources

With several events and activities scheduled throughout an academic year including travel to the spring national ACS meetings, securing sources of funding is quite critical. We have been fortunate that we can request funds from other university organizations. Two sources that have consistently given us funding upon request are the university's Council of Clubs and Student Government Association. These two sources combined provide us with ample funds to defray the cost of accomodations for two faculty advisors and 10-15 student members. Funds for travel and registration are covered in part by funds received by the university's Committee on Student Research (CSR) where one of our faculty advisors, Birgit Mellis, is a member. Student researchers can submit an application supported by their research professors to request funds for travel to a conference. Other funds to assist in travel costs come from our own fundraising activities. Our major fundraising activity is selling laboratory coats, safety goggles, and organic chemistry laboratory notebooks. These are usually sold at the beginning of each academic year. The combined funds raised from these sales and CSR funds are usually sufficient to cover registration fees and travel costs for 10-15 stdudent members.

Concluding Remarks

The success that the ACS Student Chapter at the University of St. Thomas, Houston, TX, has achieved throughout the course of its 16 years of existence has been primarily due to the commitment and dedication of the people that make up the chapter: faculty advisors, chapter officers, and members. Without those people, the chapter would not be where it is today. We hope that, as the chapter continues its work towards the future, it is able to continually inspire not only in its members, but in all people coming in contact with its representatives, the transforming power of chemistry and to advance the broader chemistry enterprise for the benefit of our planet and its inhabitants.

Acknowledgments

We gratefully acknowledge all past and present student members of the ACS Student Chapter at the University of St. Thomas, Houston, TX. We also acknowledge Thomas Malloy, emeritus professor at the University of St. Thomas, who started our chapter and was faculty advisor for the first 12 years of our chapter's existence.

References

1. www.stthom.edu.
2. www.facebook.com/ust.acs.3.
3. philambdaupsilon.org.
4. acsghs.wildapricot.org.
5. nisenet.org.
6. Ledesma, E. B.; Mullery, A. A.; Vu, J. V.; Hoang, J. N. *Ind. Eng. Chem. Res.* **2015**, *54*, 5613–5623.
7. Ledesma, E. B.; Hoang, J. N.; Solon, A. J.; Tran, M. M. H.; Nguyen, M. P.; Nguyen, H. D.; Hendrix-Doucette, T.; Vu, J. V.; Fortune, C. K.; Batamo, S. *Ind. Eng. Chem. Res.* **2014**, *53*, 12527–12536.
8. Ledesma, E. B.; Hoang, J. N.; Nguyen, Q.; Hernandez, V.; Nguyen, M. P.; Batamo, S.; Fortune, C. K. *Energy Fuels* **2013**, *27*, 6839–6846.
9. Ledesma, E. B.; Campos, C.; Cranmer, D. J.; Foytik, B. L.; Ton, M. N.; Dixon, E. A.; Chirino, C.; Batamo, S.; Roy, P. *Energy Fuels* **2013**, *27*, 868–878.
10. Alemany, L. B.; Malloy, T. B., Jr.; Nunes, M. M.; Zaibaq, N. G. *J. Mol. Struct.* **2012**, *1023*, 176–188.
11. Tran, T.; Malloy, T. B. *J. Mol. Struct.* **2010**, *970*, 66–74.
12. Alemany, L. B.; Malloy, T. B., Jr.; Nunes, M. M. *Magn. Reson. Chem.* **2010**, *48*, 882–891.
13. Gonzalez, E.; Dolino, D.; Schwartzenburg, D.; Steiger, M. A. *J. Chem. Educ.* **2015**, *92*, 557–560.
14. Joseph, S. M.; Palasota, J. A. *J. Chem. Educ.* **2001**, *78*, 1381–1383.
15. Covey, T. M.; Lopez, A. G.; Tran, A. H.; Tran, A. J.; Crawford, W. C. *J. Undergrad. Chem. Res.* **2014**, *13*, 18–20.

16. Schwartzenburg, D.; Nguyen, R.; Ishak, Y.; Sosa, H.; Gifford, A.; Mnjoyan, S.; Steiger, M. A.; Palasota, J. A. *J. Undergrad. Chem. Res.* **2009**, *8*, 71–73.

17. Ayad, M.; Kusters, I.; Schwartzenburg, D.; Palasota, J. A. *J. Undergrad. Chem. Res.* **2007**, *6*, 161–163.

18. Vu, T.; Marruffo, L.; Palasota, J. A. *J. Undergrad. Chem. Res.* **2005**, *4*, 27–29.

19. Lakhani, S.; Tinnerman, W. N.; Palasota, J. A. *J. Undergrad. Chem. Res.* **2003**, *2*, 83–86.

20. Hamzo, M. G.; Pereira, B. C.; Oviedo, R. J.; Palasota, J. A. *J. Undergrad. Chem. Res.* **2002**, *1*, 161–163.

21. Oviedo, R. J.; Pereira, B. C.; Palasota, J. A. *J. Undergrad. Chem. Res.* **2002**, *1*, 19–22.

Chapter 4

Mining the Hidden Wealth of Community College: A Guided Journey

Doris Joy D. Espiritu*

Founder and Adviser, ACS-Wright College Student Chapter Chairman,
Physical Sciences and Engineering Department, Wright College, One of the
City Colleges of Chicago, 4300 N. Narragansett Ave., Chicago, Illinois 60634
*E-mail: despiritu@ccc.edu

Community colleges are different from four year colleges and universities. Aside from the fact that most of our students only attend community college for two years, community colleges are commuter schools. Most of our students work and commute. Getting involved with college events requires more effort. Building and sustaining an award winning ACS Chapter from Wright College, one of the City Colleges of Chicago, has been a challenge and an opportunity. Understanding and meeting the demands of the diversity of student population in terms of ethnicity, socioeconomic status, academic preparedness, and social skills has been instrumental to making ACS-Wright College Student Chapter (ACS-WCSC) an award winning organization. Working the limitations of a two year college and designing Chemistry related activities focused towards students' needs and interests have paved the way to sustaining an award winning Chapter.

Wright College Facts and Statistics

Wright College is located at the northwest side of the city, and the City Colleges of Chicago is one of the nation's largest community college districts and is the largest in Illinois. The district serves more than 120,000 students seeking to enter the workforce, pursue higher education, and advance their careers in seven independently accredited Colleges. The student demographics vary from college to college. In 2015, Wright College served more than 21,500 students (13,000 credit students). Sixty two percent are minorities (51% Hispanic and 11% Black). Out of the 13000 credit students, 1,500 are pursuing a degree in natural sciences (*1*).

Serendipitous Beginning

"What are you doing at a community college? You are wasting your brain!". A resonating comment from a colleague and a friend. A comment that I took seriously as a challenge. He was wrong, but he could be right especially since his comment was precipitated by my breaking news: "I am tenured".

Teaching Chemistry at Wright College was my first exposure to community college system. I never really understood what a community college was, nor heard about the stereotypes of going to community college. I know I was hired to teach Chemistry and I love it. My friend's comment bothered me not because of the "label," but because of the possibility. Although I enjoy teaching, my work was becoming more of a routine after my tenure. I indeed felt, without the challenge of a demanding research expectation that I was slipping away. It was a scary thought, so I participated in many college committees to keep using my brain. Even with my participation, I felt there was something missing. I chose to be at community college to make a difference inside and outside the classroom and I knew I could, but I just don't know what and how at that moment.

A few weeks later, I caught a lucky break. One of my students in General Chemistry I, Jessica, approached me out of boredom. She was not bored with chemistry, but bored with being at Wright. I found out that Jessica had a Masters in Psychology. So what was she doing in my Chemistry class? Jessica pursued a degree in Psychology at one of the universities in Chicago. She did not find job after graduation, so she pursued graduate school and finished her Masters. After receiving a large sum of student loans to complete these degrees, she realized that Psychology was not really for her and ended up taking chemistry classes at Wright to pursue nursing.

By listening to Jessica, I discovered hidden treasures of Wright College, which we exploited to building ACS-Wright College Student Chapter.

Hidden Wealth 1: *Wright College has extremely talented and experienced students whose potentials and leadership skills are to be discovered and maximized.*

A typical Chemistry class at Wright College has 35 diverse students. These students are either: 1) recent high school graduates who did not get in to university of their choice; 2) recent high school graduates who are extremely bright with stellar high school records, but do not want to invest on student loans; 3) older students who are going back to school after a long time; 4) immigrants who have degrees in their countries and are retaking their classes at Wright; and/or 5) course takers who have bachelor or graduate degree who are taking Chemistry classes to prepare for career change or professional entrance exam like MCAT and PCAT. Common to most students are the facts that they work and they commute, therefore building connections among their peers are often neglected.

Jessica was bored because she did not have access to students with the same interests. On a typical day, students take classes and often they go home without talking to anyone. From listening to Jessica, I knew what I needed to do. I decided that I would build a club to provide students like her a community. A club which is broad to spark everyone's interest, but is focused on one theme. A club with a sense of purpose, that would provide networking opportunities, professional development, leadership training and community service, yet would remain focused on "Chemistry". My conversation with Jessica was the birth of ACS- Wright College Student Chapter.

While Jessica was doing research on how to start a Chemistry Club, I was looking for students who were Chemistry or Chemical Engineering majors. Unfortunately to my disappointment, there were only very few. What I found out was something more. There were non-chemistry majors who were looking to boost their resume in order to strengthen their transfer application. Most students also need to do a lot of community service to be competitive.

Hidden Wealth No. 2. *Since we have a limited number of students majoring chemistry or chemical engineering, the organization has to be built to serve both chemistry and non-chemistry majors.* *This provides diverse talents and skills.*

Hidden Wealth No. 3. *Students at a community college have to compete to transfer. A membership in an academic organization will help students build their academic resume.* *A membership to a national organization like ACS is attractive.*

Hidden wealth No. 4. *Students need community service to be competitive.*

After surveying students' interests, we decided to build not just a local Chemistry Club, but an ACS Student Chapter. Although Wright College is different from a university, I turned to Dr. Marjorie Jones, ACS- Illinois State Chapters adviser for advice. Her suggestions were valuable. While Jessica was busy recruiting possible members, I was writing an ACS Student Chapter start-up grant proposal. Together we drafted the Chapter's constitution. We started as a Wright College Chemistry Club on Fall 2011. We received monetary and materials support from ACS on December 2011, and on March 2012, Wright College was formally recognized by ACS Board to be an ACS Student Chapter Affiliate.

ACS-Wright College Student Chapter provides our students opportunity to network, lead, collaborate and serve the community through chemistry in ways they would otherwise not experience at Wright College. It certainly takes a lot of time investment to build the chapter, but as a founder and advisor, the time invested benefited not only many students, it made me grow personally.

My friends' comment and having Jessica (with Masters of Psychology) in my General Chemistry I, happened for a reason. It might be serendipity, but now, I can confidently say that "wasting my brain" at Wright College does not occur to me anymore after five years of having organized and run the chapter. I have enjoyed thoroughly guiding students' journey towards their goals as they learned the skills of running a professional organization, all of which grew their love for Chemistry while building their resume.

Sustaining an Award Winning Chapter: A Guided Journey

When we built ACS-Wright College Student Chapter, winning was not part of our agenda. We didn't even know that ACS recognizes student chapters for their activities. With research, collaboration, and monetary support from ACS and a lot of time invested, Wright College built an outstanding ACS-Student Chapter. The first activity report was submitted in congruent with ACS startup grant report. Our biggest activity in 2012 was building the organization, member recruitment and a few K-12 community service activities. ACS peer reviewers awarded us Honorable Mention for our 2012 activities. The reviewers' constructive suggestions did not go to waste. That Honorable Mention award together with reviewers' suggestions provided us motivation to keep going. In 2013, 2014 and 2015, ACS-Wright College Student Chapter were recognized for its Outstanding Chapter Activities. The Chapter also won Green Chemistry Award in 2014.

Building ACS-Wright College Student Chapter was not easy. Sustaining the Chapter was harder. Sustaining an award winning Chapter, I feared, was almost unattainable. We could have not done it without learning from our mistakes, by making our weaknesses our strengths, and by celebrating our failures and successes. Through team work, collaboration between the members and the advisers and using the suggestions from ACS Chapter advisers who spent their time reviewing our activities report, we improved. The support from our College President, Vice President, Activities Director and a few of the Chemistry faculty members and staff also boosted participation among our students.

Recognizing Our Dilemma

Fast Turnover

The biggest dilemma that caught the Chapter by surprise was the fast turnover of officers and members. Most of our students transferred one semester or one year after they joined the organization. Even Jessica who helped build the organization only stayed on her leadership role for that particular semester. I originally failed to advise officers to plan ahead because, although I saw it coming, I was privileged to work with very strong founding members. Luxury did not do our organization a favor. For instance, when Jessica stepped down, the organization was still motivated for Spring 2012 because Bridget, our Chapter's founding Vice President, stepped up to the plate and led the Chapter in 2012. Reality kicked in when Bridget and the rest of the founding officers, except the secretary, transferred in Fall 2012. Our Chapter was left as a skeleton.

Funding

Getting students to join meetings, especially organizational meetings, needs funds. We initially provide these students with food and other resources from our Student Government Club budget. The number one challenge for running a student organization at Wright College was that we were no longer allowed to use the

Chapter's funds for food and other marketing materials. With funding difficulty, we struggled to promote our Chapter. The logic of not allowing the organization to use its own funds is incomprehensible, but we have no choice but to follow regulations. This discouraged some of our members from participating in fund raising activities and posed a huge challenge for the Chapter.

Low Interest in Chemistry or Chemical Engineering in the College

As previously mentioned, out of the 1500 hundred students majoring in natural sciences per year, only a handful are interested in Chemistry and chemical engineering.

Marketing and Documentation of Activities

The hardest part of maintaining an award winning organization is soliciting participation and marketing events. Proper documentation of events is also important and requires a lot of time investment. Students' schedules are not synchronized with each other, and most student members are hard to reach outside of the classroom. It is the job of the President to document or at least collect documentations for activity report, but the president is not present at every event.

Working through Our Limitations

ACS-Wright College Student Chapter is a young Chapter. We started with less than 20 members five years ago. We are now boasting close to 80 active members, 26 of those are ACS National members, and we have one ACS scholar. Even with our dilemma, our members thrive and we continue to grow. How did we sustain our chapter through these difficulties and continue the winning tradition?

Fast Turnover: Lessons Learned and Emergency Measures

I missed a very important aspect of advising an organization. I neglected to advise my student leaders to always plan ahead. Our founding officers were great, but they were all gone by our second year. Our Chapter had to reorganize and emergency measures had to be taken. Instead of electing officers, we had no choice but to appoint a President and a Membership Director in Fall 2012. Michael, a chemical engineer major and a pioneer student member, agreed to serve as President for one semester. The President and the Membership Director together with the veteran secretary rejuvenated ACS-Wright College Student Chapter. Our mistake was almost a deadly experience, but we learned from it. To prevent the same situation from happening again, we designed *continuity* measures.

1. Our officers have to be intact before the start of each new academic year. They serve as the anchor of the Chapter's recruitment activities, especially during turnover. All Officers are to be elected in March of

the previous academic year and shadow with the existing officers. They develop from team building and leadership skills trainings, especially on recruitment and membership retention. They are also required to be a National ACS members. ACS-Wright College Student Chapter pays the Officers' National ACS membership and renewal dues as an incentive.

2. Our Membership Director plays a salient role for keeping the Chapter going. We recently organized our most successful membership recruitment strategy, a "Member recruitment and retention competition." This competition recruits members by providing our new members mentorship partners. As an incentive, the member who recruits the most new members receives a monetary award during End of the Year Celebration.

3. We hold fun activities for members during the summer and winter break to keep the members connected and to reenergize the team. We do bowling, billiards, laser tag, and get together parties.

4. We maintain visibility throughout the year. This includes classroom visits on the first few days of classes, collaboration with student government, updating our kiosk in the College lobby, and promoting our organization through a student constructed website and social media site.

Investing time on these measures has been very successful on sustaining our Chapters activities. In fact, although students are not always on campus during the summer, two of our major community service activities, "Back 2 School Program" and "Science Volunteer and Demo at Illinois State Fair" are well attended. These measures alleviated and solved our existing problem on fast membership turnover.

Funding: Planning Ahead

ACS-Wright College Student Chapters train students leaderships skills especially in handling a very difficult issue: Funding problems. Our leadership learned to plan ahead. We plan strategies to get funding for food, marketing, and other activities. Sometimes we do potluck meals. Although it is not acceptable that the students cannot use the funds they raised for their activities easily, planning months ahead developed problem solving skills and saved us from a lot of frustrations. Our students also developed Deal-Making strategies. An important skill not just for a Chemistry organization but for life.

Non-Chemistry Major Leadership Positions

Ideally, an ACS Student Chapter should be led by a student with interest in Chemistry, but Wright College students are mostly medical, engineering, or healthcare majors. Fortunately, Chemistry is a central to medicine, engineering, and other healthcare fields. We continuously design activities that spark interest for non-chemistry major. We also offer leadership skills and networking opportunities to attract student leaders. Sustaining an award winning Chapter is capitalizing on each student strengths and maximizing their potentials while exposing them to the field of Chemistry. For example, our web developer and community service director for fall 2015 is a computer science major. He does not have to be a chemist to recruit members and design our website but he plays a major role on sustaining the success of the Chapter. In return, he went to the 251st ACS National Meeting in San Diego and learned about Computers in Chemistry. Since we were founded five years ago, ACS-Wright College Student Chapter were able to influence at least five student members to major in Chemistry. A few of them graduated with a degree of Bachelor of Science in Chemistry.

To sustain an award winning ACS Chapter in a community college with a very few students majoring in Chemistry, we exploited the diverse interests and talents, hone students' skills for our members and the Chapter's advantage while doing Chemistry.

Technology and Communication Strategies

It was originally the President's job to ultimately document all activities but it became a group effort now. Our President, our Secretary, and our Membership Director/web designer have access to ACS report page. The rest of our officers and members have access to ACS google drive. Members can upload their documentations anytime. In terms of marketing, most officers are in "Group me app". Each officers are charged of mentoring at least 4 other members of their reach. A particular officer is responsible for reminding members about meetings and activities that they would otherwise not attend because they are not opening their emails. We also keep our Kiosk and bulletin board at the Physical Sciences and Engineering department current.

Typical Chapter Activities

Although we tailor our activities to our members' needs and interests, we have developed a set of events to do every year. Most of these activities include K-12 outreaches, community service activities, and team work and leadership trainings. The two events listed below are pre-semester community service events we used to kick off the academic year. We also held leadership and team work activities before the academic year starts.

Back 2 School Distribution Event

This is a pre-semester event we do every year. Together with Illinois Currency Exchange, our ACS-Wright College Chapter plays a role in inspiring K-8 students to go back to school each fall with enthusiasm and a love for learning. Our role is to perform science/chemistry demos to entertain and educate K-8 kids. We performed at least six 20 minute demos to a total of 500 kids. Figure 1 shows curious and excited kids in one of the Back2School demonstration events.

Figure 1. Back 2 School distribution event. ACS-Wright College Student Chapter entertaining and developing curiosity among kids. The kids on this picture were looking at a color change demo using dry ice. (Photo courtesy of Doris Espiritu).

Figure 2. ACS-Wright College Student Chapter College at Illinois State Fair, Springfield Illinois. (Photo Courtesy of Doris Espiritu and Eduardo Perez).

Illinois State Fair

The state-wide Illinois Fair is another pre-semester summer activity. Figure 2 shows ACS-Wright College Student Chapter participated and promoted science during the Illinois State Fair. This volunteer opportunity involved travel to Springfield, Illinois and interaction with ACS- Local Chicago. This event provides networking with professional and meeting a lot of people in Illinois while promoting Wright College and ACS- Wright College Student Chapter and Chemistry.

Academic Year Activities

1. *Fundraising*
   We only do two fundraisings a year, a safety glasses sale and a bake sale. The safety glasses sale is our biggest fund raising event. This happens at the start of every semester. Chapter members do not only sell safety glasses, we also visit chemistry classrooms and talk about the Chapter, recruit members, and chemistry tutors.
2. *Chemistry Week*
   Chemistry week is the busiest time of the year for our Chapter. The Chapter usually plan a week long activity that sometimes extend through the weekend and the following week. We designed activities to celebrate Chemistry Week according to the Chemistry Week's theme. We hold the bake sale during this week. It has been customary that we do a big Chemistry demo show open to Wright College community. A Chemistry quiz is usually the highlight of Chemistry Week, and we celebrate Mole Day at a local grade school showcasing Chemistry. We also hold a separate award ceremony for the Chemistry quiz champion on another day.
3. *Fun Activities*
   We hold Halloween and Holiday Parties for all members and alumni to get together. We usually hold this outside the campus so members can bring guests. Members engage in activities like laser tag, bowling, and billiards. These events developed team work and friendship among present members and alumni.
4. *K-12 activities and other community service*
   From November to January, our members as a group volunteers at local grade schools to judge science fairs. In February, we also design, run, and judge a Science Olympiad Chemistry lab event. In March, our members as a group judge "You Be the Chemist". Throughout the year, we offer free chemistry tutoring to the whole Wright College community.
5. *Earth Day and Green Chemistry*
   In collaboration with Environmental Club and Student Government Association, we celebrate Earth Day. Our activities vary.
6. *ACS- Local Chicago Section Interactions and National Interactions*
   Our biggest interaction with ACS Local Chicago Section is through our volunteer work at the Illinois State Fair, but we also attend dinners and

meetings locally. Nationally we support the Young Chemist Society's Chemistry in a Box webinar. Depending on the topic, we facilitate the webinar at Wright College which is open to the public. We also send four to six students every year to ACS National Meeting.

7. *Professional Development*
   We hold professional development days. Our topic ranges from team building activities, leadership and soft skills training, demo practices, and even laboratory techniques like titration skills enhancement.

8. *General membership and officers meeting.*
   The officers meet two times a month and the members meet once a month to plan for activities. The rest receive regular communications through email and social media.

9. *Appreciation/Recognition and Awarding*
   We recognize our members and officers who perform above and beyond to make ACS-Wright College Student Chapter a success. We hold two recognition events per academic year.

Looking at the Bigger Picture

Wright College, like most community colleges, is different from four-year colleges is many ways. The length of time students stay at a community college is shorter. Consequently, we face a high membership and officer turnover rate. Students also commute which makes it harder for students to stay on campus to participate in college activities and events. Even with these limitations, Wright College is full of enough hidden wealth to sustain an award winning ACS student chapter. Our very diverse student population is certainly an asset. As founder and advisor of ACS-Wright College Student Chapter, I see nothing but opportunities. The events we hold throughout the semesters provide our members opportunities to serve the community and to develop themselves as future leaders while learning Chemistry. Each ACS student member has unique needs. Constantly designing events including leadership training, field trips, community service, and networking opportunities that suit our members' needs are always part of the Chapter's agenda. Interesting activities which are conceptualized and planned by officers and members play a vital role in advancing future opportunities. Team work and participation produce success and success produces pride. Our day to day achievements facilitate the Chapter's ownership that slowly form bonds between members of Wright College and the community in general. Our Chapter members separate themselves from the rest of the students because, through our activities, our members get to know more of our faculty, improve their networking, and thus provide themselves opportunities for building their resume.

Sustaining an award winning ACS Student Chapter starts with capitalizing on our very diverse student population. Our diverse student talent, academic skills, interests, socioeconomic background, and racial and ethnic profile make our chapter vibrant and dynamic.

The core requirement for chapter success semester after semester is managing our fast turnover of members. Our *continuity* measures provide us success so far.

Other Factors

Adviser

Wright College provides students a more personalized attention. As founder and adviser, building and sustaining an ACS Student Chapter are investments of time. This investment is already paying off by the quality of student leaders we produced from ACS-Wright College Student Chapter for the last five years. Most of our student leaders transferred and sought out the similar opportunities at their four-year institutions. Often they ended up reactivating a University Chapter or wanting to start one. The advisers' time and commitment are very important, especially in the community college where our turnover is really fast. Guiding the new set of officers is salient to the Chapter's growth. I believe that passion is contagious. Making the members realize that the adviser is passionate not just with the organization, but especially to their individual success, has developed student's passion for the organization. I witness the facts through ACS-Wright College Student Chapter.

We sustain an award winning Chapter even with fast turnover because our members are passionate in what they do.

Alumni Connection

A bonus to being an ACS-Wright College Student Chapter members are our alumni who are either done or already transferred to universities. A few of our members attend our holiday party and speak to our members how ACS- Wright College Student Chapter provided them opportunities.

References

1.   Facts and Statistics, City Colleges of Chicago. http://www.ccc.edu/colleges/ wright/menu/Pages/Consumer-Information.aspx (March 24, 2016).

Chapter 5

Sowing the Seeds of Chemistry through Student Chapters: A Journey Full of Commitment, Enthusiasm and Passion

Ingrid Montes-González*

Department of Chemistry, University of Puerto Rico, Río Piedras Campus,
PO Box 70377, San Juan, Puerto Rico 00936-8377
*E-mail: Ingrid.montes@upr.edu

This chapter presents the successful, unique and/or common activities of the Student Chapter of the University of Puerto Rico, Río Piedras Campus. Each year since 1993, this Student Chapter has been rated by the ACS as Outstanding with the exception of one year in which it was rated Commendable. In this chapter the activities are organized according to the categories from the Chapter annual reports (e.g. community service, NCW, CCED, professional development, Chapter development, social functions, proposals, communication, and others). The chapter also includes examples of successful activities that could help in being designated as a Green Chapter.

Introduction

The Student Chapter of the University of Puerto Rico, Río Piedras Campus each year since 1993, has been rated Outstanding with the exception of one year in, which was rated Commendable. We are grateful and delighted for this opportunity to contribute this chapter to share our experiences.

ACS student members are exposed to experiences that enrich, strengthen and develop extra skills beyond their academic formation (*1*). For example, leadership and determination are key qualities that students may not develop in a classroom or a laboratory setting.

One of the most satisfying feelings for a professor is to see student members evolve into great model citizens, rational, and giving human beings. This is why community service is so strongly emphasized in this chapter. The importance of community service is highlighted through the coordination of activities that give members the chance to develop empathy towards others as well as making the community service part of their mission in life. Moreover, it is a way to promote a greater understanding of chemistry concepts that are presented during demonstrations. It could also inspire creativity and spark interest in the teaching-learning process, as in our case, with the publication of a demonstration based on a solubility concept (2). Student members also gain knowledge in important notions, such as Green Chemistry, and obtain a more rounded formation about environmental issues and consciousness on how anthropogenic factors could be managed, to mention only a few.

Millennial students are described as the ones that expect instant gratification. It is true that student members obtain a great fulfillment after each intervention, but is not instant, as it requires a lot of work and commitment. This is helpful because the chapter develops in them organizational skills, as they have to devote extra time on each project, simultaneously fulfilling their academic responsibilities.

Another important aspect of the Chapter experiences is networking. Maintaining close contact, communication, and network with the Local Section and the other Student Chapters is fundamental. Teamwork is a very important skill that could be developed in student members. This is particularly easy nowadays with social media and virtual communication. Puerto Ricans have historically been characterized because of our social and jovial spirit. Throughout the years we have fostered interactions and team-work between our Chapters, each one being an essential part of one body, the Puerto Rico Local Section. As one body we have to work in concert and towards the same goal.

The following sections, are intended to share some of the activities that seem to be the most successful, unique and/or common that could spark others creativity in the development of a robust and healthy Student Chapter.

Service

Service is one of the main goals of our Chapter. This section is organized according to the different activities that target school students as well as other members of the general community. The activities that will be addressed are: visits to school, interaction with ChemClubs, National Chemistry Week, Chemists Celebrate Earth Day and finally other community projects.

Visits to Schools

Each year the Student Chapter visits many schools. The purpose is to informally educate students in elementary, middle, and high school to develop an interest in science, especially chemistry, and to encourage them to pursue studies in chemistry related fields. In order to fulfill this objective regular visits to different schools are carried out. Different chemical reactions are demonstrated in order to reveal how chemistry is involved in everything that goes on the students' regular lives. We mostly utilize household materials in order to help them associate the chemical concepts to their daily activities. We call our demonstrations "The Magic of Chemistry", show because when the younger children see our demonstrations and the chemical reactions that form part of our performance, they might at first believe it like magic. We present different demonstrations covering topics like acid-base reactions, density, pressure, polymers, solubility, redox reactions, and luminescence. After we finish each demonstration we give them an explanation on the chemistry and its relevance to everyday life.

Aside from "The Magic of Chemistry" shows, we also perform hands-on demonstrations at different schools, especially at the elementary level. During these demonstrations, we encourage children to actively work different hands-on activities involving several household materials in order to show them how chemistry is involved in their houses and in their bodies, and to awaken their interest in science. It is important to mention that in both types of demonstrations we always emphasize safety rules.

Interaction with ChemClubs

Throughout our visits to schools we interact with ACS ChemClubs, that are established in high/secondary school (*3*). The organization and foundation of ACS ChemClubs throughout the high schools of Puerto Rico has became one of our major goals. We are focusing in the organization of activities that involve devoted high school students and their schools. In our opinion, the importance of the ChemClubs resides in the improvement on the educational development that these students undergo and help them into becoming passionate and committed professionals who can eventually turn out to be distinguished scientists. Our targeted audience is mainly composed of Hispanic high school students who are usually predisposed towards chemistry as being a difficult and boring subject. We seek out to change this perception and broaden the student's concept of what chemistry is about.

One of the ChemClubs activities that we have successfully implemented is a Biennial ChemClub Congress. This innovative activity has been implemented since 2013 as a joint effort with one of the first Puerto Rico ChemClubs (Visual Arts High School of San Juan). The goal of this activity is to provide ChemClubs from public and private schools, the opportunity to participate in a scientific meeting, full of educational activities, learning experiences, entertainment, and interactions between participants among many other benefits. Around 150 students from 12 to 15 ChemClubs and their respective Faculty Advisors participate in each Congress. Each Congress has been developed under a main theme and has integrated an art contest to be used in the promotion and during the Congress. The program includes a plenary lecture by distinguished scientists (e.g. Dr. Luis Echegoyen, Dr. Mary Kirchhoff), four to six concurrent workshops on hands-on demonstration experiences, presentation of resources from ACS (including Chemistry Olympiads, Project SEED, and Scholars Program), panel discussions executed by some ChemClubs in relevant topics such as ethics, and social interaction among participants. The first ChemClubs Congress was held in 2013 and the main theme was Nanotechnology. In 2015, we hosted the second one and the main theme was Nature and its Chemistry. Both of them were very successful and a very enriching experience because they fostered new bonds between the ACS Student Chapter members with the ChemClubs. These congresses also contributed to their academic development as high school students got exposed to new topics that were not discussed or worked in detail in their schools.

In addition, our Chapter members gained from the Congress because during the coordination, they developed leadership skills and became role models, and while helping in the workshops, they acquired teaching experiences. We are very proud to say that these ChemClubs have become so enthusiastic with their mission that they have taken part in our most important activities, such as the celebration of National Chemistry Week (NCW) and Chemists Celebrate Earth Day (CCED).

Other successful activities that have been implemented are: training for chemistry demonstrations, participation in our schools' visits, including them as protagonists during the chemical demonstrations (after receiving training), sharing effort to develop some contests during the academic year, and serving as mentors for Scientific Fairs.

National Chemistry Week (NCW)

In order to make NCW (*4*) a success, meetings with Chapter members are held on Saturdays and weekdays after-hours within the months of September and October to plan, organize and train for the activities emphasizing the importance of following safety rules. The demonstrations to be presented during NCW are chosen very carefully. We work really hard practicing and discussing the chosen activities for each event. Every volunteer is required to understand and be able to explain in a simple manner the chemistry concepts involved on each presentation and the relationship that each has with the yearly theme. Practice is a key aspect of our success; students rehearse with other peers and receive feedback on their performances. Below are some of the most successful activities.

For several years, an Open House for High School Students has been an important part of the NCW celebration, where over 500 high school students are invited to our Campus. The students have the opportunity to visit and explore different chemistry research laboratories and the core instrumentation facilities (e.g. NMR, MS, nanotechnology, etc) on our Campus. A research student (undergraduate or graduate) from each laboratory prepares a brief presentation about the research they are conducting and how chemistry is involved in every process, including the operation of the instruments. Also, they carry out hands-on activities that are prepared based on the yearly theme. The demonstrations have proven to be a success. The students are always curious about how chemistry is present in everyday life and show to be fully interested in the chemical processes that are involved in each demonstration. Last but not least, the students are gathered to join a one hour "The Magic of Chemistry" show that reinforces the main objective of the event: to illustrate once again that chemistry lies within us in every single process. The volunteer work of our student members is essential for the success of the show, the laboratory tours, and the demonstrations.

NCW coincide with the celebration of the Mole Day. For this activity a table is set up where students from our College of Natural Sciences could walk up and find interesting facts concerning Mole Day. Also, trivia questions are asked in order for them to receive gifts and snacks. During the activity the participants of the trivia are able to encounter chemistry questions, mathematical problems, concepts about the mole, etc.

There are other successful activities that are included as part of the NCW celebration, however they also could be held all yearlong. One of them are the Goofy Games. This funny and relaxing activity involves the participation of our student members, professors, and graduate students that compete. Our Goofy Games are characterized for including traditional games (trivia questions, egg races, etc.) as well as games that tackle different laboratory skills, such as assembling a distillation system.

Seminars help our Chapter members in the development of their careers and and enrich their knowledge. The Student Chapter coordinate some seminars to provide relevant information about the NCW yearly theme. Furthermore, every year, the Chapter plans and organize several seminars which integrate different areas of interest for our members. Some of them are: cutting edge research, graduate school opportunities, summer research experiences, chemistry job opportunities, the chemist profession, how to prepare a good presentation, opportunities in research, and different new technologies available nowadays. Members are also welcome to attend seminars coordinated by the different Departments and programs of the College of Natural Sciences.

With the advancement of technology, webminars are also very popular where students receive the same benefits in a virtual fashion. We promote participation of our members on webinars sponsored by ACS. In the case of the new initiative of the Spanish webinar series, we encourage our members to participate. To facilitate interaction among members, we deliver them in an amphitheatre, as a regular seminar and provoke discussion on the theme.

Our NCW program always includes some other activities that are directed to high and middle school students. One example is to work with Boys and Girls Scout to earn their chemistry badge. As described before, visits to schools are very important for us. Because the number of students who can participate in the Open House event is limited, we also include some visit to schools as part of the NCW celebration. Depending on the yearly theme, we also include exhibitions in our Library as part of the NCW activities (e.g. Puerto Rico Water Authority regarding the water sources and treatment in the island, nutrition, astronomy, among others).

"Festival de Química"

Our Faculty Advisor is the founder of the outreach event called "Festival de la Química" that recently was adopted as an ACS Program (Festival Series) (5). Through this event she promotes interactions among all Student Chapters and ChemClubs in Puerto Rico. The " Festival de Química" is always part of NCW and CCED celebrations and it has become a tradition among the ACS Student Chapters from Puerto Rico. It is usually coordinated as the opening NCW activity. It joins the efforts, not only of the ACS Student Chapters, but also the ChemClubs, and other sister Societies such as as the Caribbean Division of the American Association for the Advancement of Science (AAAS), and the Puerto Rico Chemists Association, among other professional organizations, depending on the yearly theme. The number of volunteers has been increasing and during the last years more than 500 volunteers participated to welcome thousands of kids and the general public. Each group of volunteers is responsible for presenting demonstrations related to the yearly theme and for emphasizing the importance and relevance of chemistry in daily life. Some of the groups enrich the event through dramas, music, and songs. The Festival is open to everyone and the public has the opportunity to participate and interact in the chemical demonstrations, conducting hands-on experiments. Also, a positive family involvement is promoted.

Chemists Celebrate Earth Day (CCED)

The traditional and already described "Festival de Química" is always part of the Chemists Celebrate Earth Day celebration (6). As explained before, it joins efforts from all Student Chapters in Puerto Rico as well as some ChemClubs. The yearly theme is always emphasized through the presented activities. Besides this initial activity, events similar to those of NCW are sponsored visit to schools are held during the course of the week to integrate the student community to our celebration and, most of all, to educate and create an environment-friendly conscience in the public. Other unique successful activities will be described in the following paragraphs.

As part of CCED celebration, a survey was conducted to determine the level of general knowledge concerning actual environmental problems. This survey consisted of ten simple questions in which the students from the College of Natural Sciencies anonymously expressed their knowledge and concern for the environment. The results were published in the Student Chapter's monthly newsletter. This issue emphasized CCED included articles related to environmental problems and provided students' perspective for simple solutions. It also included games, puzzles and chemistry jokes.

Collaboration with Other Student Organizations is very important for us, so the Student Chapter always promote interaction with other College of Natural Sciences student organizations. One example is the collaboration with the Eco-environmental Society (SEA). SEA organized a bohemian night where students from the whole campus participated. We helped them with this big event by setting up a hands-on demonstration that permitted us to educate the public about the environment's great importance and also mention different easy ways to stop contributing to Global Warming. We emphasized on the importance of reducing our carbon footprint.

A kite Competition was another successful activity during the CCED celebration. The goal of this activity was to promote casual gathering of the undergraduate and graduate students in a learning environment. The students who attended were asked to mention the principles of Green Chemistry and/or other environment friendly advices as a token for their coffee and pastries. The different laboratories were informed about the kite competition and were asked to build a unique kite in order to compete.

One past CCED yearly theme included activities related to "Reduce Your Carbon Footprint". During that year a special activity was coordinated where members visited the General Chemistry academic laboratories giving freshmen students a talk on global warming, greenhouse effects and the necessary reduction of carbon emissions. After the presentation we gave the students our now famous "green tips or advices" where we taught them how simple modifications in our daily life can improve Earth's health and how making a commitment to apply these "tips" would benefit us all in the long run.

Some workshops were organized for another year where the theme was related to Climate Science. These workshops were coordinated in collaboration with the Local Section and aimed to enhance understanding of Climate Science through student members and high school students (ChemClubs), moreover, lead them to become Chemistry Ambassadors for Climate Science. The workshops included hands-on activities, tips on how to work with children/students, and information on those chemistry topics related to Climate Science that are typically included in the science curriculum. The workshops were based on the resources already available at the ACS web page, including the Climate Science tool-kit. Furthermore, we added some hands-on demonstrations to facilitate the understanding of the concepts. We provided resources/materials to each Student Chapter/ChemClub (brochures, Powerpoint presentations, and some household materials needed for the demonstrations). These trainings used the "train-the-trainer" model. It was expected that each participating Student Chapter or ChemClub would develop at least two activities to promote public interest in Climate Science issues and to improve its understanding.

Other Community Service Projects

As mentioned earlier, one of our Chapter's main objectives is to enhance the overall undergraduate experience to provide a learning environment that goes beyond the classroom. We recognize that each student must not only grow at the academic level but also need to be developed in all areas that define an educated individual, a model citizen, and a rational human being. For this purpose, one of our main areas of emphasis is community service. We highlight the importance of community service through the coordination of different activities that aside from the scientific and academic responsibilities give the members an opportunity to develop and enrich their empathy towards others and understand the importance of giving back to the community. Among the activities that have been successfully implemented are:

Earthquake and Hurricanes: Before, During, and After

One of the Student Chapter's more important guiding principles is its concern for the community. The Chapter organized a workshop of what to do before, during, and after natural disasters such as hurricanes or earthquakes. This workshop was coordinated in collaboration with the American Red Cross and the Department of Environmental Science of our College of Natural Sciences as well as local volunteers. Student members had the opportunity to train people from the general community in safety standards and precautions that should be taken.

Orchard Project

This was an innovative initiative directed towards community service. The orchard is located at a low income community near our Campus. The target audience of the project were children and youths who daily attend and work at the orchard after school. This young community ranged from elementary to high school students. The goal of the Orchard was to keep these children and youths away from the streets, providing them with tasks and responsibilities concerning the maintenance and development of the orchard. The goal of our Chapter's project was to integrate knowledge and motivation through the wonderful tool of science into the Orchard's project. We shared our "Magic of Chemistry" show one afternoon and a couple of hands-on demonstrations to illustrate different chemistry concepts. This project was a brand new experience where the student members dedicated their time and skills to bond, motivate, and inspire the youth of this community.

Visit to "Virgilio Dávila Day Care Home"

The Virgilio Dávila housing project is a high-risk area. The volunteer program provides shelter to battered children from 4 to 12 years old. The student chapter organized a recollection of food and first need articles. We visited them and worked with them, as well as played some games. We also presented some chemical demonstrations.

Make a Difference Day

This activity was held in La Rosaleda, a low income housing project. The ACS UPR-RP Chapter collaborated with many other recognized student organizations in Puerto Rico, as well as the American Red Cross Association. All of these associations cooperated to create a fun, engaging, educational activity for children and their parents. It was great to see the result of a joint effort to give these kids a different day and more importantly, to motivate them to study and continue in school.

Recollection of Toys and First Necesity Items

Our members got involved in this recollection drive and afterwards donated the articles to shelters, Make a Wish Foundation or homeless people near our Institution.

Donations for Haiti

Due to the devastating earthquake that took place in the sister country of Haiti on January 12, 2010, leaving thousands of people homeless and in crisis, the student members in coordination with the Department of Chemistry unified efforts in order to raise funds for the victims.

Visit to the Boys and Girls Club, Senior Citizen Homecare, Children's Shelters, Hospitals, and Shelter for People with Special Needs

Our visits have given us an enormous opportunity to start an awareness of the social problems in our community as well as the chance to explore and expand our sensibility towards our fellow human beings. With these visits we obtain a satisfaction that goes well beyond any academic achievement, we get a sense of fulfillment and happiness that has nothing to do with being a chemist, but a well rounded person. We feel that we are part of our community and that we are trying to make a better world and do our community a greater good.

Sponsor of Pediatric Cancer Patients Foundation

"Un Rayito de Sol en tu Habitación" is a non-profit organization that helps pediatric cancer patients and their families through the treatment process and also support in dealing with loss. This foundation works at the San Jorge's Children Hospital, where cancer investigations are currently being held. We collected items of personal needs for the children and also accompanied them during a "Three Kings Day" feast as part of the Holiday season, where we had the chance of getting to know the pediatric patients and their families, while sharing our hands-on demonstrations with them. Some of our members have felt so connected and engaged with this initiative, that they have joined this foundation as volunteers.

Volunteers for the Children Museum

Each year, in November, we share our demonstrations with ChemClubs and the general community at the "El Morro Castle" at San Juan Puerto Rico during the National Children Day. This is a premier National Park and visitors attraction. Other similar initiatives to this one are the closing activity of the Bio Alliance Week and the Bristol-Myers Squibb Open Houses and Family Days.

Chemistry and the Art Workshop at the Ritz Carlton Hotel Kids Club

This initiative was developed by one of our members who work with children at this hotel. In this workshop we worked with children as small as 3 years up to 12 years old. We helped the kids understand the close relationship between Chemistry and Art. This activity was a 4 hour workshop in which the kids learned how to make clay, flowers, butterflies, slime, and more materials that they use in their everyday activities. We helped them establish an early connection between chemistry and art that will hopefully remain in their memories. This workshop was a great experience to integrate areas that may seem completely unrelated to people. It gave us a great opportunity to apply our knowledge and teaching skills to explain chemistry concepts to kids.

Mentors and Tutors for Diverse Courses and/or Laboratories

This initiative is personal for each member involved and demonstrates their true commitment towards their peers. Being a tutor is not easy; these students must keep theirselves up to date with the topics covered in the course for which they are tutoring without falling behind in their own classes and responsibilities. Yet they find time from their own busy schedule to teach others what they have learned in class.

Safety Week

In this week, chapter members visit General Chemistry academic laboratories and offer a series of demonstrations in order to explain the different safety rules that must be followed in a chemistry laboratory. Many of the students taking General Chemistry are freshmen students; at the end of the presentation the chapter members offer information about the ACS student chapter, its activities and how they can join if interested.

In conclusion, our Chapter provides a high quality service to our Department and University throughout the academic year. Thanks to our trajectory we are often requested to represent at our College of Natural Sciences, our Campus and the University of Puerto Rico System in different activities inside and outside our Campus (e.g. EXPO at UPR System level, Bioscience Week, Research Week, University Open Houses, summer camps among others). We have the privilege of being recognized as the best (amongst all the others) Student Organization in our Campus for being the most active and recognized Student Organization. This makes us very proud and motivates us to continue to work hard to support our Campus and all of its students. Working and giving our service for the University is a way of saying thank you for the high quality education received.

Professional Development

One of our goal is to help our members to develop as a well rounded professional. exposing them to different experiences were they can learn about their profession and maintain themselves up to date with novel findings in diverse areas of chemistry and science in general. For this purpose, we have coordinated or participated in visits to industry, workshops, fieldtrips, seminars, webminars, and plenary lectures as well as casual discussion forums in the form of Science Cafes coordinated by the Local Section.

We have coordinated an Ethics Forum at our Campus, where controversial ethics problems and important scientific issues were discussed by our members. We provoke a casual atmosphere, having donuts and coffee for the attendees to encourage their participation.

As part of the ACS, the largest scientific organization, we are committed to encourage our members to participate in scientific meetings in order to enrich their academic and professional formation and expose them to networking. The attendance to local, regional, and in a very special way, ACS National Meetings, provides the students with an incomparable networking experience where they can share their research project results and gain insight from other fellow students and scientists. The experiences acquired in scientific meetings open their minds on all the possibilities for graduate school, internships, and jobs. These types of participation provide the student with a networking and a clear preview of what a researcher's world is comprised of. Furthermore, they realize the resources that ACS provides and benefits of being a member.

We continuously encourage our members to join undergraduate research programs. Many of our members are part of federally funded programs, such as MARC (Maximizing Access to Research Careers), RISE (Research Initiative for Scientific Enhancement), PRLSAMP (Puerto Rico Louis Stokes Alliance for Minority Participation), and NASA undergraduate fellowships, among others. Both, in scientific meetings and research programs, our students are encouraged to take in summer internships outside of Puerto Rico. This offers them a chance to do research in fields that may not be available at our Campus. Additionally, summer internships in another institution provide a wider scope in the student's research at their home institution. Furthermore, this exposes them to greater cultural diversity and gives them an opportunity to explore different visions of university life outside of the Island.

There are many other ways that Chapters could foster professional development, one of the most important one is to collaborate with the Departments to host especial events as it was the case for ACS on Campus. Our Department of Chemistry hosted the first ACS on Campus held in Puerto Rico. Our Chapter was part of the organizing committee and members worked eagerly in the coordination and promotion of the event. We worked closely with them to ensure that the coordination fulfilled ACS and the speakers standards. It was a full day event that included relevant presentations related to ethics, safety, careers, publications, and entrepreneurship. All participants evaluated the activity as an enriching experience.

Another huge opportunity has been to collaborate with the Local Section in hosting a Regional Meeting (SERMACS 2009). This ACS regional meeting was held for the first time in Puerto Rico. We worked eagerly as the host for the undergraduate program. The program included talks by distinguished scientists, workshops (teachers, undergraduates, and K-12 students), socials events, a "Festival de Química", and scientific presentations by our members. This meeting gave the opportunity to attend two Nobel Laureate Conferences. This was a true honor for all because this is not your typical scenario to have a Nobel Laureate Conference live in your town. During the meeting we also hosted a students' visit to our Campus facilities in benefit of all the attendees that came from other places and were interested in what our Campus has to offer. This activity was coordinated with the Chemist Graduate Student Organization. Besides the general Campus, visits to different laboratories in our facilities were arranged in order to motivate students to pursue graduate studies at our institution.

Our Student Chapter is always willing to participate and collaborate in special celebrations, as was the International Year of Chemistry (IYC) in 2011. During that year the Chapter supported and collaborated in all the activities hosted by the Local Section. Some of the activites were: five "Festivales de Química", the Global Water Experiment, ten Science Cafes, Seminars, as well as working very hard with other professional associations to ensure the success of the 2011 IUPAC World Chemistry Congress that was held in Puerto Rico.

Chapter Development

Recruiting new members and retaining our members have been the most important goal over the years. The earlier the student gets involved with the Chapter the higher the probabilities that they will become leaders. Typically, a student joins the chapter in the second or third year of their undergraduate career, meaning that when they finally gain the experience to become a leader they might be graduating and leaving the student chapter. On the contrary, if freshmen fall in love with the Chapter they will begin their involvement in the activities and learn more about what the mission is and how it is carried out. This leaves them with three to four more years to grow as Chapter members and leaders, learning from the mistakes of past boards and contributing with new ideas and future plans for the improvement of the Chapter.

To maintain a healthy succession for our Student Chapter, it is very important to share our passion and mission with the new members that become part of our family each year. Some of the activities that have advanced this goal are presented in the next paragraphs.

Traditionally, as a recruitment strategy, we engage students from Biology, Mathematics, Physics, and Environmental Sciences. In an evolving scientific world where interdisciplinary approaches are rapidly taking over it is our duty to incorporate different points of view and perspectives into our Chapter. By involving individuals with different areas of expertise we gain an important advantage by means of the access to the knowledge of all the fields and the opportunity to incorporate them into our activities for the benefit of the whole community.

At the beginning of the academic year, we set up a booth in a centralized location of our College where board members actively recruited new members. At the booth we have different handouts available, explaining the mission of the Student Chapter and its major objectives for the year. Additionally, we have different appealing demonstrations in order to draw the attention of the students and spark their interest in the Chapters activities. Another strategy is that we request Professors' permission to make announcements in the classrooms before classes start. This allows us to reach those students who have not hear about the Chapter by other means and get connected with active members. Finally as already described, during the first week of the academic year, chapter members visit general chemistry laboratories and explain the importance of safety. However, at the end of the presentation Chapter members offer information about the ACS Student Chapter.

Once we recruit new members and to esure that the "Magic of Chemistry Shows" and the "hands-on demonstrations" will be successful, we conduct training workshops. We firmly believe that is important for the volunteers to be able to explain the demonstrations to the students according to their academic level. Futhermore, student members need to be able to verbalize their explanations in a simple way for the younger children or to explain in full detail if the participant is an adult. It is even more important to be prepared for any curious question that could arise. In order to achieve this, we provide the volunteers with the specific demonstrations along with the chemistry concepts involved in them.

Regular Meetings/Assemblies are essential for our chapter's organization, coordination and success, therefore they are held as often as necessary. In these meetings the members of the Chapter's Board, the Faculty Advisor and other members have the opportunity to get together for the coordination and planning of different events. During the meetings there is a brainstorming of ideas or the proposal of new activities to increase our members and community's interest in Chemistry. We encourage the participation of the student members in order for them to collaborate in the creative process, help throughout the coordination of the different activities and feel as an indispensable part of the Chapter. Attendance of the members to the meetings depended mostly on their academic and personal responsibilities consequently sometimes is not as high as expected.

Social Functions

Social interactions work as our best recruiting activities because they attract people from different backgrounds, majors, and Colleges in our University. Social activities also motivate student members to volunteer for future activities and most importantly help them have fun during their tough semester schedules. Our social activities are the perfect opportunity we always offer to our members to help them relax from their increasing academic workload. They present an outlet for expression and help students establish connections between undergraduate and graduate students, as well as with the Faculty and Department personnel. It is throughout these informal events where board members, student members, and faculty members have the chance to interact with each other without pressure or stress, This allows the formation of strong bonds within our members.

Our first Chapter Meeting always include a Pizza Party. Undergraduate Chemistry students participate in a meeting to talk about the Chapter goals, planning and committees. The Chapter president, faculty advisor and the Department Chair express the importance and opportunities that the Student Chapter provides. We end the meeting hosting the social pizza party to spark members interaction.

The Initiation Ceremony is always a huge success with high participation of the new members, as well as the old ones. We all have fun and become inspired to continue to work for the rest of the year. In this special event, members bring their families and we also invite professors and Faculty administrators to let them know what we do as a student organization, helping others and showing everyone the transforming power of chemistry.

At the end of the academic year we host a closing activity. This traditional activity could be a family day or any other social interaction for the members. Here we recognize their hard work. Expressing gratitude towards our members is essential since they are the key component of all activities and events. We understand that it is important to maintain our members motivated and more over to say thank you with a great party where they can share and have fun.

Other activites that are common for student organizations and that have been successful in our Campus are: special parties (e.g. Halloween, Thanksgiving, White Christmas, St. Valentines' Day), softball games (against professors, or graduate students as well as other student organizations), movie nights, etc.

Proposals Submitted to ACS-Education

The ACS Education Office provides funding opportunities in the form of small grants. These are known as Innovative Activities Grants and Community Interaction Grants. Some succesuful projects are sumarized below.

Innovative Activities Grant:

What Can I Do with a BS in Chemistry?

The major goal of this project was to organize and conduct a series of activities directed to senior students at our Campus which do not have adequate knowledge to write a resume, conduct an interview, apply for the GRE exam, apply to a graduate program or take the chemist license exam, which is a requirement in Puerto Rico. We also included a panel of speakers representing industry, government and academia to provide a forum for question and answers for students interested in understanding what kind of jobs they can pursue with a degree in chemistry.

A Gift of the Magic of Chemistry to Children's Hospitals

The major goal of this project was to organize a volunteer program and conduct a series of activities related to the "magic of chemistry" topic directed to sick children from 4 to 12 years old. The Student Chapter organized a food and first need items drive to give to the children. These activities included visits to work with them, some chemical demonstrations as well as playing games.

Community Outreach through Qualitative and Unique Interactions (C.O.Q.U.I.)

The major goal was to share Chemistry with children, regardless of their special needs or physical diversity. The project consisted in targeting five different children populations and presenting some demos adapted according to their needs to help them understand some important concepts of chemistry.

Fostering a Safety Culture

Safety rules are usually introduced within a slide show presentation that does not get the deserved and required attention. The goal was to develop a safety video in Spanish that emphasizes the exposition to potentially hazardous situations that students could encounter and provides with information on how to avoid or deal with these situations conscientiously. This project supports the Guidelines of the ACS Committee on Professional Training that establishes the importance of the emphasis on safety. Moreover, this project also supports the report of the ACS Committee on Chemical Safety that establishes the need for creating a safety culture in academic institutions.

Community Interaction Grant

Having Fun with UPR-PR Chemistry Majors

The major goal of this project was to organize and conduct a series of activities directed to high and middle school students at those schools that do not have adequate resources to provide adequate chemistry laboratory experiences.

"The Magic of Chemistry Show" Tour

This was one of the first proposals of the Student Chapter and a keystone to establish the robust ongoing program. More than 3,000 children in six different towns attended our performances that involved chemical concepts showing how chemistry is found in our daily life.

Chemistry and Art

The major goal of this activity was to organize a contest related to the NCW "Chemistry and Art" topic directed at high and middle school students. The contest was held at the Central High School, a community school specialized in Visual Arts. This activity was conducted in a very dynamic, interactive, and practical manner, in which the student worked in a piece of art at the same time that they performed an investigation about the chemistry involved in the artwork. It included painting, ceramic, artistic design, digital art, publicity art, sculpture, among others. Some well known artists as well as chemists participated as judges. The exposition and the award ceremony began with violin music played by high school students.

Green Chemistry Essay Contest

The objective of this project was to organize and conduct chemical demonstrations and hands-on activities related to "Green Chemistry" with six graders from four different schools. It also included and invitation to them and their teachers to a workshop (hands-on activities), coordination of an essay contest on green chemistry and an exposition of the work done by the students during a special award ceremony. We gave the teachers educational material related to the Green Chemistry, including the video prepared by the ACS.

Boys Scout Chemistry Badge Program

This project was already discussed as part of the NCW section. The main goal of this project was to organize a Badge Program that included activities (talks, demonstrations and hands on experiencies) directed to high and middle school students at the San Juan District. Our challenge was to ensure mastery of some specific chemistry concepts and change their perception of chemistry as something difficult, boring and of no relevance or importance.

ChemClubs in Puerto Rico

ChemClubs was also discussed in a previous section. The major goal of this project was to organize and establish ACS ChemClubs in Puerto Rico. The target audience was high school students, (100% Hispanics population). This proposal was helpful to spark the many ongoing projects explained before.

Other Grant Opportunities

ACS Division of Analytical Chemistry Grant

This special grant was offered to support the Global Water Experiment as part of the International Year of Chemistry celebration. Our Student Chapter joined efforts with ACS Puerto Rico Younger Chemists Committee to work on a program seeking the following objectives: to develop test kits for K-12 students and teachers; convey the knowledge on the importance of water purification and the different methods of analysis to the public (undergraduate, graduate, and overall community); collect local water samples and measure the pH using colored indicator solutions and a pH meter; and create our own database and contribute to the Global Experiment database.

This was a very successful activity that impacted thousands of people locally and internationally, because it was included as one of the main activities of the outreach program of the IUPAC World Congress hosted in 2011 in Puerto Rico. It was also included as part of many other events hosted by the ACS Local Section during that year.

Green Chapter Activities

Our Chapter understands and take very seriously their role as leaders and possess a genuine desire to teach their classmates about the importance of becoming more "green", not only in our academic or research laboratories, but in our lives outside campus. We believe it is very important to point out the differences between Green Chemistry and other environmental initiatives such as recycling. What makes Green Chemistry unique is its emphasis in prevention over remediation; we understand that this is the winning option for the development of a sustainable community and further vital knowledge and awareness for developing scientists. As mentioned before, webinars, are great sources of interactive information directly with the National American Chemical Society

The proposal "Going GREEN" (Generate awareness, Recycle, Educate, Energize- Natural Sciences Community) aims to determine how to promote awareness among the students of the College of Natural Sciences on how their actions affect the Earth. A survey was conducted during the second week of the 2011-2012 first semester with a group of about 80-90 students at the College of Natural Sciences. This helped to examine whether there was any relationship between the level of information of any student surveyed, and his or her concentration and academic year. After making the initial survey, a media campaign was conducted in the College of Natural Sciences. The campaign consisted of posters and flyers placed on the bulletin boards of the College,

Departments, Library and hallways. They contained information showing the effects of anthropogenic pollution, and also presented advice and quick and easy solutions for students to follow and be more responsible. The Student Chapter conducted informational activities and demonstrations related to Green Chemistry and recycling. Signs were rotated every one or two weeks, so that all students could see different signs and not the same ones all the time. A final survey, was conducted with the same students from the initial survey, to analyze the effect that the media campaign, including the activities made by the ACS, had on students. The final survey also included an open question, which asked the respondents if they modified any conduct during the period of the investigation. The results of the surveys were tabulated and analyzed and determined that the campaign had a positive effect on students.

Another fun and successful activity was a Trash Bin Decorating Contest addressing Green Chemistry concepts. This activity motivated our students in the proper disposal/recycling of waste, since the trash bins were much more noticeable.

A Green Chemistry workshop was presented to elementary and middle school students through arts hands-on activities. The main goal was to illustrate the applicability of Green chemistry in everyday life. Through interactive and fun art activities (drawing and painting), we explained the basic concepts of Green Chemistry.

With the visit of two representatives from the Green Chemistry Institute our chapter hosted a symposium and invited members of other Student Chapters from Puerto Rico. In this symposium, members and participants learned and understood the basis of the Green Chemistry protocols and how the use of these principles would contribute to the healthy development of our country. The symposium helped students and professor participants to become aware of how an unconscious use of chemicals can damage our environment and how us, as chemists, can diminish and in some cases abolish many chemical contaminations.

A Green Chemistry ChemClub Tour was coordinated. This tour consisted in visiting our high school ChemClubs and exchanging demonstrations of how chemistry can help improve the environmental damages that we humans are mainly responsible of. For the demonstrations we chose to show the students and other attendees how to make environmental friendly beauty and cleaning products from materials such a honey, lemon, milk, vinegar, baking soda, and sugar, we also illustrated the seriousness of acid rain with vinegar and calcium carbonate tablets and showed them their own personal carbon footprint through a demonstration involving food coloring and water. Through a series of questions we were able to illustrate how daily decisions such as bathing in hot water and eating large amounts of beef can contribute to the unnecessary and hazardous emission of green house gases.

Email messages were sent to members with "Green Tips of the Day". These consisted of a series of recommendations to be employed in our everyday life. Members could take the initiative and make simple daily adjustments to become greener citizens. For example, pay accounts by internet to go paperless, recycle electronics, and turn off electronic devices that are not being used.

A Science Cafe with Green Chemistry as its Main Topic was implemented. As explained before, Science Cafes at our Campus have been always very successful. The place for the Science Cafe is always an open space at our facility with patio tables and open access to the students circling the halls. We offered donuts and coffee for the attendees providing them with a comfortable and pleasant setting in which they felt at ease to express their opinions. More than an official activity this turned out to be a close gathering of members where they felt the confidence to ask questions and to exposed their points of view without fear of being pointed out or judged for their participation.

Green Fashion Shows have been coordinated between the Younger Chemists Committee (YCC) and our Student Chapter. In this activity we enjoyed the footbridge of over thirty dresses designed by members and non-members of various chapters including ChemClubs. All designs were prepared using only recycled material. Among the most common recyclable materials we find plastic bottles, paper, aluminum cans, and other materials. It was possible to present to the general public, which was composed by families, students, and teachers of the participants, the large amount of waste we produce and from which we can transform into something useful.

Other Important Aspects

In this final section, two other important aspects will be discussed: communication and finances. In terms of communication, is very important to maintain members informed on the activities. An effective way to communicate with members is by personal announcements in classes. Every chemistry major class is being taken by at least one member of the board where he/she can serve as link between the students taking that course and the Chapter. However, the means for effective communication have been changing and nowadays it is mostly electronic. It is important to always establish a blend of all of the avalialable means, such as publicity posters at bulletin boards, newsletter, emails, a web site, flyers distribution, social media (Facebook), Twitter, and text messages.

The last, but not the least aspect to mention are finances. Maintaining healthy finances is crucial for a chapter that wants to succeed across the years. It is not easy to raise funds, so this requires a lot of effort and sometimes is a time consuming task. During the academic year the Chapter sold laboratory items such as thermometers, laboratory coats, safety goggles, and molecular models. Also the Chapter has coordinated coffee breaks, pizza sales, T-shirts sales and hosted fundraisings events (eg. Maggie Moo's, Fuddruckers, Crepe Maker, Coldstone Creamery and Menchies). We also seek donations from local businesses (such as a well known megastore, soda bottling companies, and others) to sponsor the household materials used in the demonstrations or the snacks needed for our open houses or social events.

We are grateful that in many years we have also received donations to sponsor some of our National presentations from the University of Puerto Rico's President's Office, our Campus Chancellor, the Dean of the College of Natural Sciences, and/or the Department of Chemistry.

Final Remarks

To end this chapter, it is of paramount significance to recognize the hard work of all the students that have contributed over the years to these successful activities. We are very proud, thrilled and honored to also highlight the professional outcomes of the UPR-RP Student Chapter that have evolved into the main stream of our beloved profession. Through the engagement with the Student Chapter and the development of a number of important soft skills, many of our past student members are now recognized leaders of the Local Section leadership, serve in local, national and international committees and Divisions of ACS as well as in other professional associations. Cheerfully, some are now Faculty Advisors of outstanding chapters at the national level. Moreover, other past student members have raised from within the industrial sector involving their colleagues and companies in community outreach programs. Without any doubt, our Student Chapter Members have sowed the seeds of chemistry through a journey full of commitment, enthusiasm, energy, and passion that is a footprint that will last lifelong within them.

References

1. Montes, I.; Collazo, C. ACS Student Affiliates Chapters-More than Just a Chemistry Club. *J. Chem. Educ.* **2003**, *10*, 1151–1152.
2. Montes, I.; Cintron, J.; Pérez, I.; Montes-Berríos, V.; Román, S. A sticky situation: dissolving chewing gum with chocolate. *J. Chem. Educ.* **2010**, *87*, 396–397.
3. American Chemical Society, ChemClubs. https://www.acs.org/content/acs/en/education/students/highschool/chemistryclubs.html/ (accessed July 2016).
4. American Chemical Society, National Chemistry Week. https://www.acs.org/content/acs/en/education/outreach/ncw.html/ (accessed July 2016).
5. American Chemical Society, Festival de Quimica. https://www.acs.org/content/acs/en/global/international/regional/eventsglobal/festivaldequimica.html/ (accessed July 2016).
6. American Chemical Society, Chemists Celebrate Earth Day. https://www.acs.org/content/acs/en/education/outreach/cced.html/ (accessed July 2016).

Chapter 6

Establishing an Award-Winning Student Chapter To Enhance the Undergraduate Experience

Evonne A. Baldauff*

Department of Chemistry & Forensic Science, Waynesburg University,
51 W. College Street, Waynesburg, Pennsylvania 15370
*E-mail: ebaldauf@waynesburg.edu

To be considered for an award, ACS student chapters are evaluated annually on a variety of factors ranging from professional and chapter development activities to service and outreach events. Chapters must also assess their own progress and practices. This discussion does not seek to investigate the individual categories of award selection, but instead explores the ways in which a chapter can broadly evolve to becoming a highly functional and award-worthy organization that enriches the undergraduate experience. Through fostering faculty-student mentor relationships, establishing a productive community, and working unitedly to achieve designated goals, the ACS student chapter can create a culture that leads to a fruitful and rewarding experience for its members.

While deliberating the content to be addressed with regard to the Award-Winning ACS Student Chapter, it came to mind that the reader might first contemplate why time should be invested in such a venture. The Award-Winning Chapter is one that possesses a highly functioning structure capable of providing student members with experiences that enhance their undergraduate education. By participating in chapter activities, students gain content and technical knowledge, bolster their understanding of the chemical workforce, find a conduit for sharing their scientific passions with others, and develop organizational, task-management, and teamwork skills. Students who play an active role in their successful ACS chapter typically have a more balanced resume upon graduating, particularly with regard to their service and outreach experiences. Although a substantial time commitment is required of invested faculty in developing the chapter to its utmost potential, the pursuit is well worth the effort.

Before composing the following discussion on the ways in which our ACS student chapter has established a successful program, involved students were informally polled to determine their perspective. When asked "How does participating in our ACS Chapter benefit you directly", many answers, both expected and unexpected, were gathered and have been used to shape the following chapter. Assertions of practices that constitute successful programming are supported by providing the reader with examples of activities that have been integrated into our chapter. These examples are described with particular details of student and faculty responsibilities so that the reader (1) may gain an understanding of the expectations of all involved participants and (2) could attempt the activity should it prove appealing to their chapter.

1. Faculty Buy-in = Student Buy-in

The importance that faculty participation contributes to the success of the chapter cannot be underestimated. As is expected, the faculty advisor is essential to maintaining consistency, guaranteeing the quality of programming, and mentoring student leaders. Yet students assert that interacting with department faculty, both in the contexts of serving and teaching together, is one of the greatest benefits of involvement in ACS. Students want to share these experiences with faculty. They want to get to know faculty and have faculty know them. Our chapter is fortunate in that all faculty members in the chemistry department consistently attend ACS meetings and events. Admittedly the department is small, consisting of four full time faculty members and one Emeritus professor. It is not uncommon for all departmental faculty members to volunteer in large events. The value that this participation lends to the success of the activity is evident, but the impact on the students is paramount. Students watch faculty as they perform demonstrations, speak scientifically, and explain their ideas with confidence. Sharing these experiences with students opens a pathway for the exchange of knowledge and helps to foster mentoring relationships.

1.1. The Haunted Lab Tour

This event consists of over twenty concurrent Halloween-themed activities for attendees to experience. A mixture of demonstrations and hands-on experiments are organized throughout the chemistry department as a walk-through tour. The event targets families in the community with pre-K through 8th grade children; high school students and undergrads are also invited to attend. The tour is generally open for 1½ hours during a weekday evening. This event is a very large undertaking that involves the participation of nearly all student members. On numerous occasions over 100 community participants have attended. It is an extremely successful event that the students look forward to each year.

Consistently over the lifetime of this event, faculty members have participated by presenting demonstrations and/or bringing their families. Students enjoy interacting with the children of our faculty and teaching them chemistry! College students are always paired with faculty for specific demonstrations during the event to encourage engagement.

Student Leader Responsibilities:

- Determine the experiments, demos and activities to be performed.
- Assign locations to each activity.
- Catalog, prepare, purchase, and gather all required equipment and chemicals for each activity.
- Assign student volunteers to lead each event.
- Publicize the event to the campus & community.
- Document the event and monitor progress during the activities.

General Student Volunteer Responsibilities:

- Perform assigned task during the event.
- Participate in set-up and clean-up.

Faculty Advisor Responsibilities:

- Ensure that students are trained to present their activity.
- Confirm that all safety procedures are followed and that waste is disposed properly.
- Monitor all activities during the event.
- Assist with set-up and tear-down.

2. Creating a Community (*or, Forming Bonds*)

Students seek for a place in their college or university in which they fit and belong. The ACS Student Chapter acts as a connection point for students to engage in experiences with people who appreciate all aspects of chemistry at a level commensurate with their own enjoyment. To foster this environment, students must participate in activities and not simply attend meetings. The casual attendee is not likely to form meaningful relationships without shared experience. It may be initially necessary to incentivize participation through a purposeful mechanism such as a t-shirt giveaway, free food, or other rewards/prizes. However, if strong connections and friendships form it is likely that the chapter will not want for volunteers.

A unique method of promoting community is to allow the student chapter to have its own space. This occurred by chance within our department, and it yielded advantageous results. Following the completion of a renovation project, an unused adjunct office became available. The students slowly crept in and took over the space. Thus 'The ACS Office' was created. This recommendation for student space is one of the best community-building ideas that our chapter can offer. Regrettably not every department has the resources or space to make an office available to their ACS chapter. Yet if it can be accomplished, even through unconventional spaces or methods, the attempt is highly encouraged. Our once empty office is now aglow with hanging lights and adorned with ACS bumper stickers and other miscellaneous chemistry paraphernalia. While discussing the benefits of participating in the ACS student chapter with prospective students, the office is often shown during departmental tours. It has become a space for students to organize and be effective, to do homework together, and to discuss their research. It provides a space to de-stress. It has become a destination for them to look up silly chemistry jokes and generally have a good time. It is a space that builds community.

The following examples explore several of our best community-building activities. Events offered by the chapter should reflect the personality of the group. Experience reveals that students very much enjoy assigning enticing names to as many activities as seems reasonable. This helps to frame the events in a very positive manner, inferring that students will have the opportunity to participate in something unique and suited to their interests. These types of events contribute to the identity of the chapter.

2.1. Guaca'Mole' Day Fiesta

Mole Day is celebrated with Mexican food. Traditionally a quesadilla bar is provided. The chapter purchases basic items while individual members and faculty contribute supplemental dishes. Based on the chapter's current budget, the event may be open to the general campus or only the chapter members.

Student Leader Responsibilities:

- Purchase quesadilla supplies and related items.
- Ensure that guacamole is present.
- Decorate location (if desired).
- Publicize event.

General Student Volunteer Responsibilities:

- Bring side items.
- Participate in set-up and clean-up.

Faculty Advisor Responsibilities:

- Contribute food – ideally something homemade if possible!
- Hang out with students.

2.2. Thai & Tie Dye

This is an example of an event that evolved over time and reflects the individuality of the group members. Initially students came up with the idea for tie dying lab coats. The event was so successful that the following year it was extended by adding a meal component. That Thai food should be enjoyed after the lab coat tie dye was an obvious contention. Recently ending the event with dessert at the nearby home of a chemistry professor was also included. A catchy new name which incorporates the third segment has yet to be determined.

Primary Student Leader(s) Responsibilities:

- Purchase tie dye materials.
- Work with department coordinator so that students can purchase lab coats.
- Determine time and location.
- Help coordinate transportation to the restaurant.
- Publicize the event.

General Student Volunteer Responsibilities:

- Assist in clean-up.

Faculty Advisor Responsibilities:

- Attend!

3. Generate Chapter Pride

Worthwhile accomplishments are attained when students work together through their ACS chapter. When students understand the mission of both the ACS and their university, these accomplishments are contextualized and can readily help to instill a sense of pride in their progress. Further, students appreciate that being a part of a national organization carries weight. They value the fact that when participating in their ACS Student Chapter events, they are not isolated and independently toiling, but instead are part of a large network of chemists around the world striving to increase the knowledge of chemistry.

Students should feel satisfaction when they work towards a common goal and see good results. When fostering an award-winning ACS Student Chapter, quality programming that parallels the mission of the ACS and the university is an excellent benchmark for success. Successful events benefit both the activity participants and the student volunteers. Ideally participants perceive the time spent in the activity as worthwhile and useful, while chapter members' satisfaction is bolstered by a sense of accomplishment. The chapter advisor can play a vital role in ensuring program quality. Striking a balance between collaborating with students and letting them work independently is crucial, but quality should not be sacrificed. Mentoring students in this manner can help to instill a desire for high standards which should serve the student well both in the chapter and beyond. With this in mind, the establishment of a baseline set of high-quality foundational events is beneficial to maintain an effective chapter. These should be events that students enjoy, that challenge them to push themselves, and that serve a purpose. On campus and in the community, these events can become something that the chapter is "known for". They establish the identity of the group. Ideally the events are shaped by the personality of the members at the time, but the type of event, the audience served, and the purpose should remain constant. Events should also be diverse in scope and span the range of activities required by the chapter report.

It is important that students be recognized and respected for the efforts they undertake. One method that an advisor can use to effectively instill this sense of satisfaction in students is to regularly publicize their events and outcomes. This can be accomplished by having articles detailing their work published in the student newspaper, in alumni communications, and in general press releases. This takes time, but I encourage all advisors to assist in promoting the chapter. The results are worth the effort.

3.1. Providing Lab Experiences for Local Home-Schooled Students

Once a month a group of home-schooled students come to our university to receive a laboratory experience. Students are responsible for the entire process (planning, teaching, working through the experiment). The program is a great benefit to the participants as most students do not have access to a laboratory. We receive excellent feedback from the students and parents regarding the worthwhile experiences gained from the activity. The program is also an excellent experience for the chapter members as it requires them to put to use their classroom knowledge and lab technique to teach others. Students know that they will be expected to carry on this tradition and they look forward to it. They routinely encourage one another to lead the labs.

Primary Student Leader(s) Responsibilities:

- Plan the laboratory experiment.
- Prepare lab safety sheet for demonstration.
- Prep the lab (equipment and chemicals).
- Teach the pre-lab discussion, both the background information and explanation of the procedure.
- Act as facilitator during the lab, providing guidance to the other chapter volunteers and home-school participants.
- Provide the after-lab discussion.
- Clean & return equipment to proper locations.

General Student Volunteer Responsibilities:

- Assist with facilitation of the experiment.
- Participate in set-up and clean-up.

Faculty Advisor Responsibilities:

- Work with the primary student volunteer to decide upon the experiment to be taught.
- Address safety and waste concerns.
- Review/revise student-prepared lab procedure.
- Support students in set-up and clean-up.

- Assist in facilitating discussions (only if specifically requested by the student).

3.2. National Chemistry Week Activities

National Chemistry Week (NCW) is one of the most convenient activities upon which to build. Our chapter aims to serve both the campus and community, bringing awareness of chemistry, during the week. On campus three major events are hosted: the Periodic Table of Cupcakes, Chemistry Quiz Table, and the 'Put-a-Mole-on-It' campaign. Cupcakes, baked by the chapter members at a faculty member's home, are displayed to mimic the periodic table and are free to anyone entering the science hall. This is a favorite event that even draws administrators to the building. The Chemistry Quiz Table is staffed for an entire day by chapter members. If passers-by can correctly answer a chemistry-related question they win a prize. On Mole Day members pass out 'Put-a-Mole-on-It' stickers designed by the chapter as well as ACS fuzzy moles. To serve our community, the chapter collaborates with the local ACS section by volunteering for a large chemistry demonstration event at the local science center. For two full days, local schools and families attend the event. We purposefully choose a demonstration that coincides with the annual theme of NCW.

Primary Student Leaders Responsibilities:

- Purchase cupcake materials.
- Plan baking time and location.
- Plan transportation of cupcakes. Acquire many boxes!
- Determine and location and time for the Chemistry Quiz Table.
- Prepare a volunteer schedule for staffing the table.
- Prepare quiz questions and purchase prizes.
- Print mole stickers and order ACS fuzzy moles.
- Organize members to hand out mole items.
- Submit demonstration to local section.
- Organize student volunteers and travel.
- Organize and pack all necessary materials for the demonstrations.

General Student Volunteer Responsibilities:

- Bake cupcakes.
- Staff cupcake and quiz tables.
- Hand out mole items.
- Perform demonstrations with students and families.

Faculty Advisor Responsibilities:

- Help students bake, decorate, and transport cupcakes.
- Volunteer at cupcake and quiz tables.
- Volunteer at the science center alongside the students.
- Assist in any other ways needed!

3.3. ACS National Spring Meeting

For the cost of registration, attending the ACS National Meeting as a student is extremely valuable. The scale of the conference greatly impresses first time participants. Students report the value of the undergraduate-specific programming so beneficial that attendance at several of the sessions is now mandatory for our chapter. I specifically recommend attending the spring conference because chapters have the opportunity to attend the 'Successful Student Chapter Poster Session'. This is a fantastic opportunity for students to learn about activities that other student chapters have undertaken. Students can also learn from other chapters during the ChemDemo Exchange. Our chapter has gleaned many valuable ideas from this event, most recently several very worthwhile Green Chemistry demos, and I would encourage all chapters to visit this session. Further, the National Meeting provides countless opportunities for students to talk about chemistry in varied contexts. I have had students make connections with professional ACS members such as board members, councilors, staff, and even the residing president. These contacts benefitted the students greatly as they moved forward in their careers. The graduate school and networking events are also great events during which students can practice their communication skills and learn about available opportunities after graduation. Lastly, our chapter members LOVE the exhibition. It surprises me how readily they will approach vendors to learn about instrumentation or new texts that are available.

General Student Volunteer Responsibilities:

- Fundraise. Apply for ACS Undergraduate Program Travel Grant.
- Attending required events (Awards ceremony, SciMix, undergrad program events, specified number of technical talks).
- Presenting their experience at a chapter meeting.

Faculty Advisor Responsibilities:

- Arrange travel.
- Interact and hang out with students during meeting. Suggestion: Attend talks together! Take them to learn about the science that excites you!

Working with the award-winning ACS Student Chapter is an extremely rewarding venture. When students and faculty interact on a regular basis through outreach and social activities, a supportive and familial atmosphere can be developed and maintained. A successful chapter can greatly influence the culture of a department. Viewed broadly, generating a relational environment between students and faculty can lead to enriched teaching in the classroom, add substance to recruitment efforts, and afford favorable perceptions of the chemistry department in general. Yet the benefit to the students is greater still. Opportunities gained through chapter experiences such as developing programming, facilitating outreach activities, and working as a community serve to enhance their leadership skills, chemistry knowledge, and understanding of effective work habits. This personal growth cannot be understated. Providing quality programming for chapter members, the campus, and the community elicits pride which then inspires confidence. As students leave their undergraduate home, the ACS Student Chapter experience will continue to shape their next endeavors.

Chapter 7

A Guide to Building a Successful ACS Student Chapter

Daniel J. Swartling*

Tennessee Tech University, Chemistry Department, Box 5055, Cookeville, Tennessee 38505
*E-mail: dswart@tntech.edu

Building and maintaining a successful ACS student chapter takes time, effort and patience. It takes planning a yearly program of activities that forms the framework for student involvement. It is a shared governance between student officers and the advisors, but the advisor has to ensure that the chapter stays on course and reaches its goals.

Introduction

Being an advisor for a student ACS chapter is not an easy task. One often has to find ways to motivate students to perform their roles in the chapter as well as deal with interpersonal issues. It is just like being a parent: you want to give the students the freedom to grow and explore new ideas and possibilities while setting rules and boundaries to keep them from having a negative impact on themselves, others, or the chapter. The time devoted to service can be draining at times, but watching students mature into leaders and the impact the chapter can have on the department, the university and the community ultimately makes it all rewarding and worthwhile.

A key thing to remember is that students come and go, but the faculty advisor is the anchor, the rock of stability that keeps things going from year to year. Establishing a list of events that are always done by the chapter provides a framework of activities that the student members can always look forward to and build upon.

Fundamentals

One thing that helps with the year to year transition of leadership is to give each officer a binder that includes the specific duties of each office as well as notes taken from the previous people that held that position. This allows the newly elected officer to get acquainted with what is expected of them as well as "how to" instructions to get them started. This immediately instills confidence in that person as they have a better idea of what to do and what is expected of them as a chapter leader. The chapter at Tennessee Tech University has over time developed an executive committee composed of elected officers (president, vice-president, secretary, treasurer) as well as appointed committee chairs (outreach, green chemistry, fundraising, social activities). The chairs were added over time to divide up the work of running the chapter and to make sure that someone is in charge of certain categories of activities. It also expands the number of students that can gain experience in organization and leadership.

In order to maintain continuity from year to year, try if possible to make sure that all of the officers are not seniors. If all the officers graduate at the end of one academic year, it can be a challenge for the chapter to work up momentum for the following year. Ideally the vice-president should be a sophomore or junior that can learn the leadership skills needed to take over as president the following year.

Prior to each chapter meeting we have an executive committee meeting to discuss the agenda for the meeting including upcoming activities and events. A Powerpoint presentation is put together to show during the meeting. This is convenient for several reasons:

- It can be archived so future officers can look back at what was done before.
- It can be emailed to chapter members to remind them of what was discussed and what events are coming up.
- It can be emailed to students who missed the meeting and to students interested in the chapter (chemistry majors who are not chapter members) to keep them informed of what activities are being planned that they might be interested in.

Our chapter meets twice a month during each academic semester to allow student members to learn about activities being planned. That way they can pick activities they are interested in and sign up to help with those events.

To maintain interest in the chapter and to establish camaraderie in the students, it is important to plan social activities that are outside of the regular chemistry department setting. The chapter also needs to include the chemistry faculty as much as possible in order to have interaction with the students outside of the traditional classroom setting.

Fundraising and Finances

A chapter needs students willing to lead and perform their assigned duties as officers, but the chapter will not flourish until it has the funding necessary to pay for activities and events. It also pays for travel to regional and national meetings.

The chapter at Tennessee Tech University has a main fundraiser that starts at the beginning of the academic year with selling goggles and carbonless notebooks for the chemistry lab sections in our department. Throughout the year we have t-shirt sales, bake sales and other merchandise sales on campus. We also partner with local restaurants to have a fundraising night. Local restaurants in your area can be partnered with for a night where a percentage of the sales go towards the chapter.

Have contests for designing new t-shirts or other items for sale at fundraising events. One of our most popular and biggest selling t-shirts was designed by a student. It is an upside down periodic table shirt as shown in Figure 1. Being printed upside down allows the wearer to look things up on their shirt. We print up over 250 each year to sell to the students taking General Chemistry 1 just prior to their first hour exam. On the back is the ACS logo and name of the student chapter, which makes the shirt free advertising.

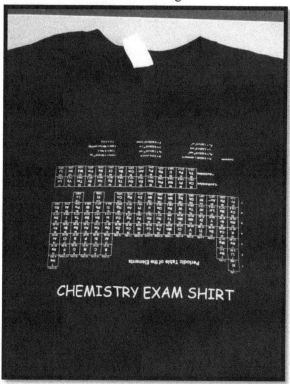

Figure 1. Upside down periodic table shirt. (see color insert)

Student government usually has a means for registered student groups on campus to receive money for events or for travel. Be sure to also check with the dean or the provost for additional sources of funding, especially if students are presenting research at a meeting, which also promotes the university.

The ACS Undergraduate Program office (*1*) has grants available for active chapters, including grants for travel, for community outreach and for new activities. More information can be found on the ACS website (*2*).

The chapter should have a bank account with the ability to write checks to pay for things. We have been successful in increasing sales at fundraising events by using Square readers (*3*) to take credit card payments. It makes for handling less cash and the payment goes directly into the chapter's bank account. And it is easier to keep track of expenses when the treasurer writes checks for things like pizza and drinks for meetings.

Social Media

One of the hardest things to do is to try and keep everyone informed and communicating with each other. The traditional means of advertising meetings is through flyers and banners posted in the chemistry department and around campus. The problem is that flyers and banners are treated as background noise and ignored as students walk around glued to their cell phones. The classic means of communication has always been the email distribution list. Students again are often not very attentive in checking their email for messages from the chapter.

One thing that has helped with communication has been a chapter Facebook page. Posting of chapter meetings and the creation of event pages has been helpful in keeping members and non-members informed of activities. An additional feature of Facebook is allowing chapter members to post pictures of activities that can be used later for the annual report. Pictures can be tagged with the names of people photographed which helps in identifying who was involved in activities.

The advisors and officers of our chapter have been using GroupMe (*4*) to rapidly keep each other informed and to ask questions and have them answered. Other means of communication include a Twitter account and a chapter Gmail account. Signup sheets for activities are available on Google Docs as well as the secretary's spreadsheet for attendance at chapter meetings and events.

Travel Contract

The chapter at Tennessee Tech University routinely sends 15 to 20 students each year to the spring ACS national meeting. Thus the bulk of our fundraising is budgeted towards travel. To make it fair to all of the students involved, we have instilled a travel contract for each student that wishes to travel to present research at a national meeting. The contract has to be signed by the student and by their research advisor. The contract requires the student to attend a certain number of chapter meeting, participate in a certain number of chapter events, and help with a certain percentage of fundraising activities. Fulfillment of the contract allows the student to be reimbursed 100% for travel costs. This keeps each person that wants to travel accountable for helping to raise the funds needed to attend national ACS meetings.

Green Chemistry Status

The ACS Green Chemistry Institute (5) grants awards to student chapters that have done at least 3 activities that encompass the 12 principles of green chemistry. Green chemistry is **not** synonymous with sustainability, ecology or environmental chemistry, although there is some overlap among the areas. The events and community outreach planned by the chapter must clearly show which principles of green chemistry are being emphasized. Chapters that qualify will receive a special award at the spring national meeting. The award is a rather unique plaque – made from laminated bamboo, which is a sustainable renewable resource.

Attending Scientific Meetings

An important means of professional development for students is to attend scientific meetings. Attending meetings and conferences allows students to gain experience in presenting talks and posters on their research, provides opportunities to network with others for career advancement, and lets the students have opportunities to socialize with their peers from other universities. The various venues are discussed in the following paragraphs.

National ACS meetings: As someone who has been responsible for undergraduate programming at an ACS national meeting (6). I can not stress enough that students that travel to the ACS national meeting should take advantage of as much of the programming directed towards them as possible. Graduate school fair, how to get into graduate school, the undergraduate banquet and mixer, the eminent scientist speaker, the undergraduate research poster session, Sci-Mix and the outstanding chapter posters, chem demo exchange, and various workshops and panel discussions are available for undergraduates to attend. Valuable information can be learned and taken back to be shared with the rest of your chapter.

Regional ACS meetings: Chapters that can not afford to attend national ACS meetings should consider attending the various regional ACS meetings. They are often of driving distance for most chapters and provides networking opportunities for students that attend. There is often special programming for undergraduates as well, such as a chemistry demo exchange, a chemistry quiz bowl, presenting research posters, and a graduate school fair. It allows for chapters in the same region to mingle and to get to know each other.

Local ACS sections: It is strongly encouraged that chapters get involved with their local ACS section. The local section can be a valuable resource for speakers and for National Chemistry Week and Earth Day events. Often times chapters will report that the drive to a local section meeting is to far for them to consider attending. If that is true, then try too host a local section meeting at your school and have the local section come to **you**. The local section may be able to provide funding for outreach events that the chapter is planning. And interacting with members of the local chapter can provide students with contacts that can help them with the next step in their careers.

State Academy of Science: One organization often overlooked by most chapters is the Academy of Science in your state. The state academy of science has poster sessions and oral presentations that are often judged and awards given. Students can gain experience by presenting their research in a smaller, more relaxed, and less intimidating setting. And as always, it provides yet another venue for networking with peers from other schools as well as faculty members and members from industry. The meetings are usually a one-day event and often on a Saturday so students do not have to miss any of their classes.

National Chemistry Week

National Chemistry Week (NCW) occurs around the third week of October each year, and is the premiere time to showcase chemistry through outreach in your community. Every year ACS has a specific theme for NCW, and it is essential that your chapter try to stick to the theme with activities and community events. For maximum impact, try to schedule one event for each day of the week. Try to have one event be a community outreach event, one event be a university outreach event, and one event be a social activity or "fun" event. The more diverse you can make things the more people will be willing to participate. An example flyer for NCW is shown in Figure 2.

Mayor's Proclamation of National Chemistry Week

One thing that can be done to publicize events for National Chemistry Week is to have the mayor of your city proclaim it to be NCW in your city and list the events that you plan to do. Since this is a local news item, your local newspaper will often take a picture of the proclamation signing and do a small article on the events that you have planned. A sample NCW proclamation is shown in Figure 3.

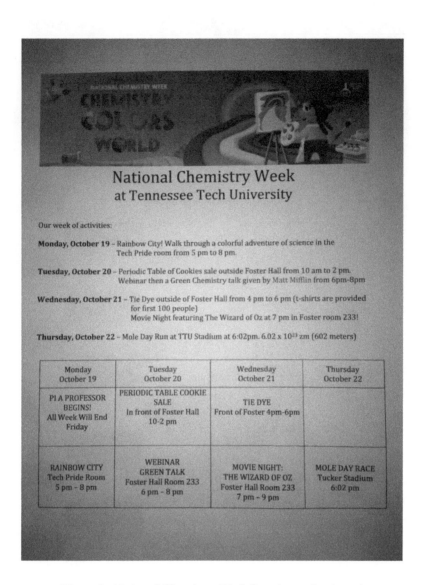

Figure 2. *National Chemistry Week flyer (see color insert)*

Figure 3. NCW proclamation for Cookeville, Tennessee. (see color insert)

Mole Day

Mole Day is traditionally observe on October 23 from 6:02am to 6:02 pm in honor of Avogadro's Number (6.02×10^{23}), the number of units of something being one mole. There are various activities that the chapter can plan for Mole Day. Printing mole stickers to place on people's backs, playing Whack-a-Mole, or having a Guac-a-Mole party are just a few examples of events that a chapter can plan. One activity that the chapter at Tennessee Tech University has planned is the Mole Day Race. Participants assemble at the university track and run 6.02×10^{23} zeptometers (zm), which is 602 meters, or about 1.5 laps around the track. There is even an official t-shirt designed for the event, which is shown in Figure 4.

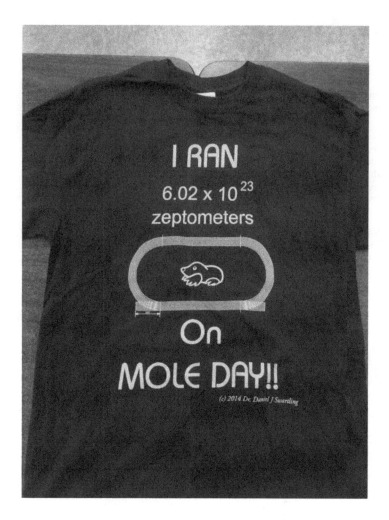

Figure 4. Mole Day Race shirt, for paticipants that run 602 meters on Mole Day. (see color insert)

Earth Day

There are many activities that a chapter can plan to coincide with Earth Day. In the past, our chapter built and maintained flower beds in front of the chemistry building, planted trees, had an Earth Day film festival, and most recently has handed out 1000 reusable water bottles with maps to campus bottle filling stations. This is a good event to also tie in a Green Chemistry activity. Going to a mall and setting up a display showing how to use greener household products is one activity that gets a lot of public attention.

Table 1. The yearly calendar of activities for the Tennessee Tech Student ACS Members Chapter.

August/September	January
•Organizational mixer •Mini-Symposium and Welcome Back Bash •Goggle and notebook sales •Fall Funfest- Stations of Imagination •1st fun activity •Fundraiser •Parent's Weekend	•Goggle and notebook sales •Merit badge university •Girl scout science day
October	**February**
•Professional Development •SERMACS •National Chemistry Week: Fundraiser, fun activity, mole day •party, mole day race, education outreach, green chemistry activity. •Halloween Party •Race for the golden helix 5k event	•Education outreach •Speaker – golden eagle chemistry seminar •Valentine Day fundraiser •Professional development •Fun activity
November	**March**
•Fundraiser •Homecoming •Fun activity •Education outreach •URECA Travel Grants •Tennessee Academy of Science	•Fun activity •Chemistry Olympiad-education outreach •Fundraiser •ACS National Spring meeting •Education outreach
December	**April**
•Professional Development •Holiday Party	•Fun Activity •Spring Awards Banquet fundraiser •Education outreach •New officer elections •Earth Day week

Annual Report

The chapter is required to submit an annual report at least once every three years in order to maintain active status. However, it is beneficial to submit a report every year because chapters can win awards based on their activities during the school year. Categories are: Outstanding, Commendable, Honorable Mention, and Certificate of Merit. It can take years of chapter growth to work up the ranks, finally winning the Outstanding Chapter Award. Once the chapter has a list of yearly events that has won it top honors, it should be easy to build upon it to maintain outstanding status. The key is to groom motivated students to take over leadership roles and to help the officers motivate the rest of the chapter members to participate in activities.

The report should be as complete and as detailed as possible. Try to have supporting documentation for each event listed. Report writing has been made easier now by the use of the chapter dashboard (7). Events and pictures/documentation can be entered as they are done. At the end of the year, when reports become due, it is simply a matter of filling in answers to questions, making sure everything is documented, and hitting the submit button. Reports are usually due in May, the results are announced in September, and the awards given out at the awards banquet at the spring ACS meeting if the chapter is attending, otherwise mailed to the chapter in late spring.

The Academic Year Timeline

The yearly framework of events is what makes the chapter run like a well oiled machine. ACS can be helpful here with Program In a Box webinars (8), as well as themed activities for National Chemistry Week (9), Mole Day (10), and Chemists Celebrate Earth Day (11). The yearly framework of events for the Tennessee Tech Student ACS Member Chapter is shown in Table 1.

Conclusion

Be sure to allow the student members of the chapter to be as creative as possible. Give the students room to try things out, to grow and to mature. But always be there as their safety net so no one gets hurt. The struggle to build a successful chapter will include some painful times, but working through the pain will lead to a strong and prosperous group. And in the end, the students will appreciate all that you have done for them.

References

1. ACS Undergraduate Programs Office. http://www.acs.org/content/acs/en/education/students/college.html (accessed 2016).
2. Student chapter grants. http://www.acs.org/StudentChapterGrants (accessed 2016).
3. Square credit card reader. https://squareup.com (accessed 2016).
4. Group Me app. https://groupme.com (accessed 2016).
5. Green Chemistry. http://www.acs.org/content/acs/en/greenchemistry.html (accessed 2016).
6. 249th ACS National Meeting, 22–26 March 2015, Denver, United States.
7. Student chapter report dashboard. https://www.studentchaptersonline.acs.org/eweb/ACSScarfTemplate.aspx?Site=ACS_Scarf&WebCode=Login (accessed 2016).
8. ACS Webinars Program in a Box. http://www.acs.org/content/acs/en/acs-webinars/program-in-a-box.html (accessed 2016).
9. National Chemistry Week. http://www.acs.org/content/acs/en/education/outreach/ncw.html (accessed 2016).

10. Mole Day Activities. http://www.acs.org/content/acs/en/education/students/highschool/chemistryclubs/activities/mole-day.html (accessed 2016).
11. Chemists Celebrate Earth Day. http://www.acs.org/content/acs/en/education/outreach/cced.html (accessed 2016).

Chapter 8

A STEM Identity Approach To Frame and Reinvent the Student Chemists Association at The College of New Jersey

Benny C. Chan,*,[1,3] J. Lynn Gazley,[2,3] Abby R. O'Connor,[1,3] and David A. Hunt[1]

[1]Department of Chemistry, The College of New Jersey, Ewing, New Jersey 08628
[2]Department of Sociology and Anthropology, The College of New Jersey, Ewing, New Jersey 08628
*E-mail: chan@tcnj.edu
[3]These authors contributed equally to the execution and writing of the chapter.

The advisors of the student American Chemical Society (ACS) members group in the Department of Chemistry at The College of New Jersey along with a faculty member of the Sociology and Anthropology Department have reframed the Student Chemists Association (SCA). Our goal is to drive the development of a Chemistry Major Identity through the STEM Identity domains of *doing, being,* and *becoming* a scientist using the activities, mentoring, and outreach events conducted by SCA. This sociological framework places the concepts such as identities, intersectionality, cultural capital, and mentoring into practice in our academic program. This novel lens offers a research-based rationale for spending a tremendous number of resources to sustain our activities, which include Happy Hours, peer mentoring, social and alumni events, seminars, field trips, fundraising, outreach, and local section collaboration. Through our examples described within the STEM Identity framework, we believe the ACS student members group at colleges and universities are in a unique position to add to the pedagogical goals of the department through transformative non-classroom experiences.

I. Introduction

The Chemistry Department at The College of New Jersey (TCNJ) has been actively working to expand the pedagogical outcomes of our chemistry majors. Through strategic planning, we have dedicated a significant number of resources to increase retention and satisfaction of our students while maintaining curricular rigor. The chemistry curriculum is typical, compared to other American Chemical Society (ACS)-accredited departments nationwide, with a focus on knowledge and skill building. We are now enhancing our program through non-classroom experiences that transform and cultivate student identities. TCNJ has been awarded several National Science Foundation (NSF) grants to address retention issues among our lowest income students where we began to understand social factors that affect performance. These best practices have refined and guided our Student Chemists Association (SCA) activities. In particular, we have begun to use theoretically guided identity development approaches to our programing to increase the academic *and* social success of our students.

II. Institutional Background and Strategic Planning

Founded in 1855, TCNJ is a public institution that has earned national recognition for its commitment to excellence. We emphasize the residential experience for its nearly 6,600 students, who are mostly New Jersey residents and benefit from a 13-to-1 student-to-faculty ratio and an average class size of 21 students. TCNJ has a freshman to sophomore retention rate of 95% and a six-year graduation rate of 87%—rates that are among the highest in the country. A strong liberal arts core forms the foundation for programs offered through TCNJ's seven academic schools: Arts and Communication; Business; Education; Engineering; Humanities and Social Sciences; Nursing, Health, and Exercise Science; and Science. TCNJ students matriculate with their chosen major and academic school with opportunities to switch schools/majors. TCNJ is committed to offering the highest quality education while remaining an affordable public institution with some of the following national accolades: the top North public school in the US News and World Report in 2015, a 2015 Princeton Review "Best Value", featured in the 2016 Fiske Guide to College, and Kiplinger's #1 best value Public College in New Jersey (*1*).

Among its many academic accolades, in 2015 TCNJ was awarded the first "Campus-wide Award for Undergraduate Research Accomplishment" by the Council on Undergraduate Research (CUR), based on their Characteristics of Excellence in Undergraduate Research. The TCNJ campus community promotes and supports a scholarly culture focused on undergraduate research and student engagement via initiatives including: faculty-student research seminars, an annual College-wide conference to celebrate student achievement, an annually published *Journal of Student Scholarship*, and an eight-week coordinated institution-wide Mentored Undergraduate Summer Experience (MUSE), in which students work side-by-side with a faculty mentor. In the nine years since its inception, over 300 STEM undergraduates have conducted full-time mentored research in the MUSE program. Likewise, of the 1,100 School of Science undergraduates per year,

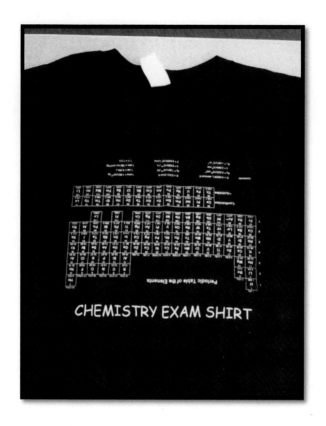

Figure 1. Upside down periodic table shirt.

National Chemistry Week
at Tennessee Tech University

Our week of activities:

Monday, October 19 – Rainbow City! Walk through a colorful adventure of science in the Tech Pride room from 5 pm to 8 pm.

Tuesday, October 20 – Periodic Table of Cookies sale outside Foster Hall from 10 am to 2 pm. Webinar then a Green Chemistry talk given by Matt Mifflin from 6pm-8pm

Wednesday, October 21 – Tie Dye outside of Foster Hall from 4 pm to 6 pm (t-shirts are provided for first 100 people)
Movie Night featuring The Wizard of Oz at 7 pm in Foster room 233!

Thursday, October 22 – Mole Day Run at TTU Stadium at 6:02pm. 6.02×10^{23} zm (602 meters)

Monday October 19	Tuesday October 20	Wednesday October 21	Thursday October 22
PI A PROFESSOR BEGINS! All Week Will End Friday	PERIODIC TABLE COOKIE SALE In front of Foster Hall 10-2 pm	TIE DYE Front of Foster 4pm-6pm	
RAINBOW CITY Tech Pride Room 5 pm – 8 pm	WEBINAR GREEN TALK Foster Hall Room 233 6 pm – 8 pm	MOVIE NIGHT: THE WIZARD OF OZ Foster Hall Room 233 7 pm – 9 pm	MOLE DAY RACE Tucker Stadium 6:02 pm

Figure 2. National Chemistry Week flyer

Figure 3. NCW proclamation for Cookeville, Tennessee.

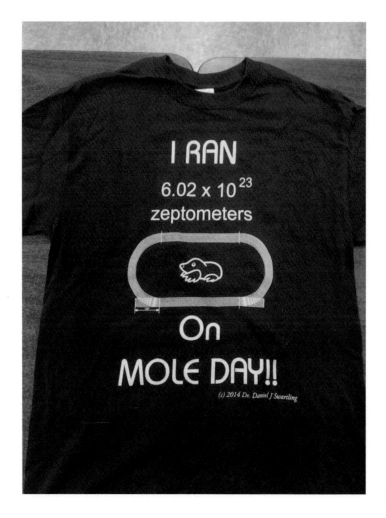

Figure 4. Mole Day Race shirt, for paticipants that run 602 meters on Mole Day.

~33% have participated in credit-bearing independent research courses at TCNJ (~75% of graduating students in Chemistry), for which faculty receive teaching credit. The School of Science at TCNJ provides funds for research supplies and to defray costs for students and faculty to present research and novel pedagogies at national and international conferences.

The TCNJ Chemistry Department provides ACS certified and non-ACS BS degrees in chemistry, with specializations in forensics, secondary education, and condensed matter (a combination of physics, materials, and chemistry with an emphasis on research). As of 2016, the Department has 12 tenure track faculty, 2 full-time non-tenure track 1-year instructors, 3 staff members (program coordinator, instrument coordinator, and stockroom manager), and ~8 adjunct course instructors. Over 50% of our full time faculty and 1-year instructors are women. The Department houses ~130 chemistry majors and graduated 34 students in 2015 (typically 30-45/year), with 26 degrees being ACS certified. Over 55% of our graduates in the last 4 years have been women, which makes any gender differences contextually similar to a biology department (currently ~58% at TCNJ) and were not explored in this work. The underrepresented minority population (African American/Black, Hispanic, and Native American) has grown from less than 10% in the mid 2000's to ~19% from 2009-15 (range 16-30%) and closely matches the changing demographics in New Jersey (2). Low income students populate about a third of our incoming class (range 24-40%, 2009-15, as measured by incoming students who filed their FAFSA with a calculated EFC, expected family contribution, of less than $12,000). Of the low income students, over 68% are eligible for Pell grants. Chemistry students come from a range of family income and cultural backgrounds.

The curriculum is typical of most ACS certified degree programs and focuses on knowledge, laboratory engagement, skill building, and research. Students pursue science careers, with approximately 30-60% attending graduate school in the chemical sciences, 10-30% obtaining industrial or teaching jobs, and 10-30% pursuing professional degrees in medicine, pharmacy, nursing, and dentistry. The Department has a 94.4% four-year graduation rate (entering students 2009 to 2012, range 86%-100%). About a third of incoming chemistry students do leave the major, but others enter the major during sophomore or junior year at almost the same rate. The Chemistry Department four-year graduation rate surpasses the overall six-year TCNJ graduation rate (87%) indicating the success of our program.

We believe the secrets to our overall success are engaged faculty, students, and staff with a rich research-based culture that invokes a wonderful holistic mentoring of students and their Student Chemistry Association (SCA). After a hiatus, SCA was reformed as a campus club in 2005, shortly after Dr. David Hunt joined the faculty. The faculty rapidly changed and expanded (7 new hires since 2005) due to retirements and the student chemistry population doubled from ~15-20 students/year in the mid 2000's to ~30-45 students/year, currently. With the new energy in the Department, SCA was positioned to increase its activities and begin to target national recognition for their activities. Our graduation exit surveys suggest SCA contributes to building the community that underpins the department's success. From 2013-16, over 70% of the students agreed or strongly agreed that SCA provided great interactions with other majors (overall response rate >70%). The qualititative student responses illustrate the community that we formed:

> "I liked the community that TCNJ chemistry's department has. We are all very close with faculty and our classmates."
> "Amazing experience. Great faculty, department was so helpful in my success here."
> "Literally the best time of my educational career, including pre-school."
> "I really enjoyed the people and environment at TCNJ and I think it truly fostered my love of chemistry, the coursework, research, and professors have been invaluable and allowed me to fulfill my current career goals."

The TCNJ campus has undergone extensive strategic planning. The 2012-2016 plan and current drafts of the 2016-2021 plan (3), include diversity, inclusion, and retention of faculty, staff, and students as major priorities. Faculty on campus have been actively working towards these goals. For example, the Chemistry and Biology Departments were awarded two National Science Foundation S-STEM grants to provide scholarships and academic support for low income students (4, 5). TCNJ was awarded an IUSE grant to study the theoretical and sociological mechanisms that drive our retention activities in collaboration with sociologist, Dr. Lynn Gazley (6). Our understanding of STEM Identity development has perfectly framed the activities of our Student Chemists Association. Our increasingly diverse student body requires us to support students as they integrate a variety of student identities with their budding Chemistry Major Identity, or sense of self as a successful chemistry major. This deliberate cultivation of a commonly held chemistry major identity among our diverse student population creates a strong sense of community in our Department and justifies the tremendous amount of resources spent on a sustained, award winning student chapter.

III. Theoretical Foundations of Our Current SCA

The partnership with Dr. Lynn Gazley has allowed the SCA faculty advisors to understand how social systems affect student success. While programs that work only on shaping the student to fit current institutional models are more common, programs that challenge existing institutions and structures are more often successful (7). Redefining the Chemistry Department's issues, problems to solve, and solutions in organizational rather than individual terms lead to better outcomes for students. For example, feeling isolated can lead underrepresented students to leave the major (8). SCA tackles these types of problems via a two-pronged approach: 1) peer-mentoring (an individualized solution) and 2) creating an integrative culture of community in the Chemistry Department (an organizational solution).

III.A. Identities and Intersectionality

The SCA attempts to help students create links and resolve conflicts between their social identities and their developing Chemistry Major Identity. Scholars argue that individuals construct their identities through an iterative interaction with social and material contexts (9–12). In these contexts, students gain and deploy resources such as knowledge, skills, and relationships in order to see themselves as, and be seen by others as, particular selves. Identity may be usefully conceptualized as a process, rather than a fixed point. For example, as they move through the major, our students develop pre-professional identities consonant with their individual goals. Further, identities are intersectional (13). Race, gender, and other social identities do not operate independently; rather, they are lived and experienced jointly. Social identities can be a model for faculty to understand concepts of implicit bias and privileges. Social identities come with *contingencies,* or the "conditions you have to deal with in a setting in order to function in it" (14). SCA is a mechanism for faculty to understand different student social identities and experiences to help address identity contingencies and institutional barriers to increase success of our students.

Social identities also intersect with academic and science identities; students navigate these intersections through self-definition and resilience strategies. Prior research suggests that a robust science identity plays an important role in science persistence, particularly when taking into account the ways that gender, race, and ethnicity can shape students' engagement with and aspirations toward science (8, 9, 15, 16). In their study of underrepresented students in a post-baccalaureate program, Gazley *et al.* argue that science identity comprises three related domains: affinity with the practices of doing science, the sense of self-as-scientist (or being a scientist), and career aspirations (becoming a scientist) (17). This offers a flexible, sensitive framework to understand the variations among these elements, how they intersect with other social identities, and how challenges due to social contingencies may ebb and flow throughout an undergraduate chemistry degree.

Using the science identity framework, we developed more specific understandings of the *doing, being,* and *becoming* elements pertinent to chemistry. The SCA first helps to ground the student within the major via developing a Chemistry Major Identity rooted in shared practices, ways of being, and aspirations. With the realization that identities will flow as career paths crystallize, we expect and support identity differentiation over time. Students may develop a Chemistry Identity if they decide to attend a chemistry graduate program or obtain a job in chemical industry. Other identities that SCA fosters are pre-health/pre-med and educator identities. The SCA focuses on the Chemistry Major Idenitity as a grounded central academic identity emphasizing the *doing* and *being* elements that helps students reach any long term goals (*becoming*).

III.B. Cultural Capital – The Contextual Resources Fueling Identity Development

Combining attention to cultural capital with identity offers a robust analysis framework to understand student-program interactions (*17*). A cultural capital perspective (*18*) argues that individuals navigate social systems using a combination of economic capital, social capital (connections and relationships), and cultural capital (skills, knowledge, and ways of being). However, how we use this capital – our dispositions, or habitus – matters as much as what cultural capital we have. Though cultural capital has been used extensively to inform program development, science programs in general rarely move beyond attention to filling perceived gaps in student skills and knowledge and frequently ignore the value of cultural capital that does not match the generally white middle-class norms of primarily white institutions (PWIs). The most successful programs using a cultural capital approach: 1) recognize and value the multiple repertoires of cultural capital new students already have, 2) address standards of evaluation within institutions, and 3) support resilience by providing safe practice spaces for students to exercise new skills (or new uses of old skills) (*19*). Taking this one step further, by imagining cultural capital as resources for identity development, programs can create appropriate context- and individual- level interventions to support student development along a continuum, moving from questions of "what do I know?" to "who am I?" SCA activities, particularly the Happy Hours (see below), were specifically designed to cultivate cultural capital and provide students with a forum where they can be recognized - and recognize themselves – as Chemistry Majors.

III.C. Mentoring

Prescriptive and developmental mentoring of undergraduates plays several key roles in student development, from the cultivation of cultural and social capital to supporting student identity work. Research finds students vary in their access to formal and informal mentors, but students with the least resources are most likely to benefit from mentor relationships (*20*). Our SCA explicitly addresses potential inequities in access to mentors by systematically employing both faculty and peer mentoring within formal and informal activities. All students receive a faculty advisor in the Department who helps to navigate the TCNJ curriculum and provides invaluable career advice; this is the intial mentor for academic success. In addition, the interactions with SCA provide peer mentors beginning the first year. Finally, if a student participates in research, the research mentor becomes an extremely rich source of cultural capital, both towards science research and career trajectories.

These faculty and peers discuss with students their academic and social goals at the college and link students with opportunities in which to apply cultural capital; without such opportunities cultural capital is worthless (*21*). However, students benefit most from mentoring actions aligned with their individual aspirations and understandings of how the world works (or schemas), which, in turn, "should be linked to knowledge of the real world and its information and resources" (*22*). Thus, mentoring works best when mentors understand students' goals, but students' goals are achievable within the context. For example, a critical role for mentoring arises as students test ideas about their future professions (the *becoming* piece of a science identity). Our students who pursue Ph.D. programs typically have participated in mentored research and discussed their choices for programs with faculty to find those that suit their abilities and career choices. In addition, faculty have assisted students to find industrial positions through personal connections with the local industry and scientific temp agencies (a primary employer for our students in the current economy), and in most instances these jobs turn into permanent positions. Faculty and peers support health career-bound students with chemistry major progression advising along with opportunities for pre-health career discussions and volunteering at various local hospitals and healthcare providers.

IV. Specific Examples of Activities

Many of our activities are common best practices that we have learned by having tremendous institutional support for faculty and student travel to the National ACS meetings ($500-$1000 per student, 5-20 student attendees/year; 1 conference/year for faculty) to present at the Successful Student Chapters at the Sci-Mix poster session along with their division-specific research posters. The Department's approach to SCA is unique through the rigorous, theory-based methodology to our activities. We believe any chemistry department can use this theoretical framework as a research-based rationale on why student chapters are critically important to the overall education process and should have significant resources dedicated towards their ACS student member chapters.

IV.A. Happy Hours

From our ACS chapter report evaluations, the Happy Hours were noted to be a unique activity conducted by SCA. The Happy Hours were designed by faculty to connect first year students to our upperclass students and faculty in a less formal environment that is more typical to networking events held at a National ACS meeting. Thus, new students gain opportunities to observe the behaviors and interactions between successful chemistry majors and faculty, and to gain practice themselves in a low stakes environment. In addition, these events support both new students and upperclass students in seeing themselves as chemistry majors and gaining recognition by faculty, which are key mechanisms supporting identity development. We designed the Happy Hour based upon the best practices from our NSF activities to increase student academic success (*4, 5*); we believe that *all* students would benefit from being given academic advice to do well in their classes. The Happy Hour is an approximately 1-hour informal biweekly meeting during the fall semester and is initially a required component for the Orientation to Chemistry course in the chemistry curriculum (CHE099, 1 hour/week for 7 weeks). All chemistry students must take CHE099 to graduate from the major and this course serves as an invaluable tool to help students successfully transition to

TCNJ. Some topics covered in a formal classroom setting during CHE099 include study habits, registration, safety, scientific ethics, along with an introduction to the library and ACS (23). The Happy Hours are organized by the CHE099 instructors and SCA executive board to bridge the academic and student social divides. When first developed, the faculty led the Happy Hour disucssions. As the years progressed, the students that had originally participated in Happy Hour found the program to be responsible to their success and began to run the discussions on their own; faculty only needed to supply the topics for discussions. Some topics covered and activities during the Happy Hours include: getting to know your general chemistry professor, academic support and study habits, a green chemistry scavenger hunt, and an introduction to research in the Chemistry Department. The only cost beyond faculty and student volunteer time for the event is to provide snacks for the meetings (~$30/gathering). The Happy Hours support students' Chemistry Major Identity in the *becoming*, *doing*, and *being* domains to help them ground their primary identity as Chemistry Majors.

IV.A.1. Students Build Cultural Capital on Being a Chemistry Major, the Chemistry Major Identity

The Happy Hour serves as a formal mechanism created by the Department to cultivate cultural capital for all students. We have found that the rigors and expectations of the chemistry major come as a surprise to most college students, particularly first generation and underrepresented students who do not have someone in their social network that knows these unspoken expectations. By cultivating cultural capital around the performance expectations and strategies to meet them, we aim to remove barriers to the *being* and *becoming* components of their Chemistry Major Identity by focusing on the specific behaviors (or *doing*) to be a strong chemistry major. In addition, this exercise required the Department to make these expectations explicit, which can have positive effects on inclusion and diversity (7).

For example, when informally interviewing students, we discovered that entering students spent only 5-10 hours per week studying in high school for *all their courses*, while maximizing time spent on extracurricular activities. During the Happy Hour, we have upperclass students discuss how much time they spend studying, which is typically ~7-12 hours/week outside of class and lab *per course*. Many of our highest academically performing students have made choices to target a small number of extracurriculars (1-2 activities) and leadership within these organizations, as opposed to the high school mentality of doing as many as possible. Our initial advice prioritizes academics as many students come with lofty goals of medical school or top 10 graduate programs that require consistent study habits and high GPAs. One strategy we have employed is to have the first year students in CHE099 put together a weekly schedule which encompasses all activities, classes, study time, and other items (eating, sleeping, hygiene, employment, family obligations, hanging out with friends, etc.) in a 24-hour period. When students start to consider the number of hours required for academic courses, they can make informed choices to best manage time between their academics and extracurricular activities. During Happy Hour, students discuss their own difficult choices between studying and non-academic activities. This exercise provides 1) new information pertinent to the context, 2) opportunities to see how other successful students have applied this information, and 3) a safe place to experiment with applying it in their own lives.

IV.A.2. Students Meet Faculty Members

Early in the fall semester, we ask that all General Chemistry professors, including adjuncts, come to a Happy Hour to meet the first year class. This serves to remove some of the uncertainty or concerns of visiting a faculty member during office hours and encourages students to see faculty as more than just classroom teachers. Faculty members should be perceived as mentors to assist in their studies and support them in achieving their goals. For many college students, the first interaction with a college faculty member is after doing poorly on a quiz, exam, or lab and this experience can be perceived as negative or "in trouble" like a high school experience of going to the principal's office or a teacher's conference. By meeting the students in a less formal environment, instructors show their human side and that they recognize the students as belonging within this challenging chemistry program. We encourage faculty members to discuss strategies they have seen or practiced as successful students in order to cultivate the students' cultural capital around study skills, and different research interests and career paths. We want students to feel a sense of belonging in our Department and introducing faculty early is a critically important component so students are more likely to attend office hours on a regular basis and develop long-term mentoring relationships.

IV.A.3. Group and One-on-One Peer Mentoring

As many faculty know, students do not always listen to our suggestions. The biggest advantage of using the student club to cultivate cultural capital is that students tend to listen to their peers. In addition, as is common in Chemistry and other sciences, our student body is more diverse than our faculty. Thus, students have opportunities to interact with students who have navigated a wide variety of social identities as they intersect with a chemistry major identity. The Happy Hour is a formal mechanism for group peer mentoring of first year students by successful students who are fully engaged in the Department. Having peer mentors who openly discuss what they do and how they think about *being* a chemistry major is extraordinarily important to the retention and Chemistry Major Identity development of our students.

SCA organizes informal one-on-one peer mentoring where the upperclass student is matched with the first year students. The summer before matriculation, SCA sends a survey to first year students to find out their interests and then upperclass students choose to mentor a student based on similar interests (either career trajectories or activity). The success can be mixed and is directly correlated to the frequency of meetings. To address this issue, our next step is to formalize the training of peer mentors to discuss concepts of diversity, privilege, and cultural capital, while requiring a minimum number of meetings between the peer mentor and mentee.

IV.A.4. Students Are Introduced to Faculty-Student Research

Faculty-student collaboration on scholarship is a Signature Experience at TCNJ and transforms students into scholars, i.e. *being* a chemistry researcher. New to the 2015 fall semester Happy Hour series was a successful event in which first year students were introduced to research experiences in our Department. This activity was employed because a common critique from chemistry majors at TCNJ was exposure to research opportunities too late in the major. To resolve this, we developed a new program in which junior and senior chemistry majors (with a variety of career trajectories from industry to graduate and health programs), currently participating in research in the different labs in the Department, attended a Happy Hour. The students discussed the research matching program used to facilitate students to join labs in the Department, what research is like in the different labs, experiences in research external to TCNJ, such as REUs and internships, and how research aids with different career goals. Another activity that introduces first year students to research is attendance at the annual Departmental poster session, held in the fall semester. The upperclass students that present their posters demonstrate their scientific skills (*doing*) and ways of thinking (*being*) to the first year students. By introducing these experiences in their first year, students can look forward to their degrees and the array of career possibilities (*becoming*). Informally, we have found that nearly all students wanted a research experience after the poster session and research discussions. Historically, 80-90% of our majors participate in undergraduate research for at least one semester, most of whom participate in three full course units of research.

IV.A.5. Using SCA Happy Hours To Understand Students from Different Backgrounds

Taking a cue from our social scientists, Happy Hours can serve as an informal focus group to understand our students better by addressing our implicit biases. Faculty may unintentionally create biases in their classrooms as we tend to apply educational practices from our own experiences, cultures, and socioeconomic status; we may be unaware of the experiences of students from different backgrounds and how these experiences may affect academic performance. During the Happy Hours, faculty would pose the questions to the first year students. Students raised the concern while upperclass students would respond to the concern first and faculty would supplement the information based on historical answers. The discussion prompts were developed through the best practices of our NSF grants (*4–6*) including:

1. What would you do differently this semester (or exam) if you could have a chance to go back? How are you going to make those changes for the next semester (or exam)?
2. What is the best piece of advice you would give to someone taking ____ class?
3. What caused you to struggle this past semester and how did you adapt?
4. What other academic or social commitments did you have during your week? How were you able to prioritize (or balance) academics?
5. What do you plan to do during the weekend (or break) and how are you going to schedule time for academics?

By building the trust with the students and having SCA peers from a wide variety of backgrounds, many first year students are willing to share more personal information in the group session while others share privately one on one with the faculty. We found that some of the common issues affect students from lower socioeconomic classes or minority groups more than those from the majority groups. For example, one of the previously unspoken expectations by STEM faculty at TCNJ is that students should remain on campus to fully engage with their academics; we are not a "suitcase campus." Some students from low income families need to work extended hours on the weekends to pay for their education. Some family cultures, particularly the first generation college students' families (a majority of whom in New Jersey belong to underrepresented groups) do not understand the TCNJ culture and expect students to be home every weekend for family obligations. Almost all of our students have commented that working too much or going home on weekends is counterproductive to their academic studies. The student must resolve the conflict, but may be completely unaware of the expectation. At the Happy Hours, the peer mentors and the faculty who have seen many of the transition issues have offered suggestions that include: lock themselves in a room to study if they absolutely must go home, ask family to come visit the campus one weekend a month, keep phone calls from home or visits at home short and sweet, find a campus work-study job that they can study while working such as the library checkout desk or the key check desk in the dorms, or find a job that is close enough to walk so they do not have to buy a car and pay for insurance. We have talked to parents at the request of students, but a more institutional solution that we hope to implement is to have a parent orientation where we can tell parents about the academic expectations and other cultural norms we expect of our majors, and to enroll the families in supporting the students. This approach makes cultural capital about expectations and strategies to meet those expectations explicit to all students (and to the faculty), creating a more level playing field. In addition, by sharing the expectations with students' families, we support students in crafting individualized strategies that draw on that crucial source of social support and cultural capital.

Bridging the culture gap between faculty and students to understand the multitude of social factors that affect student performance is important to improve the Department outcomes to increase inclusivity and the diversity of our students. The informal nature of the Happy Hour allows students to give oral and interactive feedback about their academic *and* social experience, which does not occur in our standard chemistry courses, student feedback forms, or senior graduation exit surveys.

IV.B. Beyond Happy Hour: Student Socials, Faculty-Student Meals, Collaboration with Local Section, Recruitment/Succession Planning, Seminars

SCA has developed student-only programming to build upon the success of faculty-led Happy Hours. By supporting social events, we enhance a Chemistry Major Identity that *being* a chemistry major transcends the academic realm into the student life realm, which we have found our most successful chemistry majors have done. Students have gathered for mural paintings, board/card games, laser tag, paintball, rock climbing, and take out food stress buster events. A unique formal dance, the Chemi-Formal has been an extremely popular event in which students dress up in formal wear, have a full meal and an evening of dancing and music supplied by a DJ. SCA funding and a small fee per person covers all costs for the formal. By having faculty stay away from these types of events, students are welcome to form their own communities and discuss navigation of TCNJ life in a faculty-free zone.

SCA organizes two faculty-student potluck meals each year to increase bonding and community among students and faculty. The Chemistry Thanksgiving is held in the Department during the fall semester, while the spring event is a picnic held at a local park. Holding the picnic at a local park is important to help remove students and faculty from the structured campus environment for the afternoon. SCA uses their funding to purchase food (~$150-$200) like chicken marsala, hot dogs, veggie burgers, and hamburgers to guarantee a substantial meal, while faculty and students sign up for other items. Our faculty bring crowd pleasing and interesting dishes to the events such as pulled pork, egg rolls, dumplings, buffalo chicken dip, homemade desserts, and various pasta dishes. The upperclass students collaborate with first year students to drive to the grocery store and bring a variety of food ranging from pre-washed salads and bags of chips to elaborate homemade dishes like truffles, cakes, and grilled BBQ chicken. These events create a sense of belonging in the Department with a family-type atmosphere.

In addition, SCA collaborates with the Trenton Local Section of the ACS. The president of the SCA attends local section board meetings to present the activities of the club. In addition, the local section helps support the club with some funding and volunteer support. One successful event held between the local section and SCA in 2015 was a trip to the River Horse Brewing Company. This collaboration brought together local section members with students in a low key setting. The local section members were able to network with the students to discuss careers in a low stress setting and learned about beer brewing. This event was invaluable to the students as it provided a venue to network with professionals. This again supports the further cultivation of a Chemistry Identity *becoming* by envisioning what they could become as chemists in the real world after they graduate

Through ideas gained at the ACS Leadership Institute, we have designed succession planning and recruitment of volunteers in SCA. By requiring attendance at the first few Happy Hours, we monitor the first year students that continue to come to SCA events after the required Happy Hours. These students are cultivating their Chemistry Major Identity to be a primary identity and feel connected to the Department. These students are actively recruited to become peer mentors, activity coordinators, and future leaders within SCA. One strategy that has worked well is to solicit suggestions for new activities that will encourage wider participation by students. Students that propose activities are then tasked to talk to the faculty advisors and senior members on the feasibility of the event and the planning approach. As a team, a timeline is developed and support is given to the students through regular, timely feedback mechanisms to ensure the deadlines are met. These leadership abilities are readily transferrable to any career path, as large goals must be broken down into smaller achievable actions. The Chemi-Formal is an example of an activity that was developed by student input and execution with careful advisor guidance. The students were unaware that an event of this magnitude would take 4-6 weeks of planning with a cost of approximately $1000 for food, DJ rental, and decorations. Tickets were sold to cover the cost of food and DJ rental while the club paid about $100 for the decorations. Students were glad they could work on the event slowly instead of the week before the event, which was held at the busiest time of the semester. Current organizers are instructed to always train and mentor one to two younger members to organize the Chemi-Formal the following year.

SCA sponsors a robust invited colloquium series with the support of the TCNJ Chemistry faculty and School of Science. The seminars add to the students' sense of belonging within the wider chemistry community and offer opportunities to develop in the domain of *being* of a Chemistry Identity by exposing them to larger research problems or issues being tackled at major research institutions, chemical industry, and at ACS. In 2015-16 academic year, SCA hosted two ACS Webinar in a box programs, "Chemistry on the silver screen" and "Speaking Simply." Academic speakers included: Dr. Dan Mindiola from the University of Pennsylvania on organometallic chemistry, Dr. Paramjit Arora from New York University on protein chemistry, Dr. Gerrik Lindberg from Northern Arizona University on computational fuel cell modeling, and a series of TCNJ faculty mini talks. Being in New Jersey offers our students a great opportunity to hear from speakers from industry including Dr. Nicole Morozowich from 3M and Dr. Robert Gambogi from Johnson and Johnson Consumer Products. These seminars are critical to the development of the students' Chemistry Major Identities and increase the scholarly culture of the Department.

IV.C. Outreach Activities

SCA, like every ACS student group, organizes and participates in campus and community outreach events using the great resources at ACS. While the standard chemistry curriculum focuses on developing skills in *doing* chemistry and our activities connecting students with working professionals and faculty encourage students to refine their goals toward *becoming* professionals, outreach activities support students in the identity domain of *being* a chemist or chemistry major. While research in chemistry gives students a sense of *being* a chemist, the activies occur later in the curriculum at TCNJ. Outreach events, a form of service learning, offer opportunities for students to be recognized, and recognize themselves, as competent chemists and chemistry majors *early* in the curriculum without adding to the already packed chemistry content. In addition, such outreach activities can support students in developing an identity rooted in giving back to the community, which can be an important factor in persistence among underrepresented students (*9*). Some examples of our outreach events include giving out periodic table cupcakes in the dining hall, green chemistry posters on catalysis and the principles of green chemistry, National Chemistry Week themed events like candy, elementary school science fair judging, and demos like dry ice bubbles, high school outreach that include research talks, and a campuswide week of science, which included a demonstration using spectroscopes. In these outreach events, the students are the experts in chemistry, the scientists that the audience can ask questions and engage in discussions about their experiences. Students gain skills to confidently speak about science (and present themselves as scientists) in front of a lay audience, which has been shown to increase persistence in and identification with the Chemistry Major is critically important to their development as a scientist (*24*).

TCNJ strongly values service-learning as a Signature Experience and received a Carnegie Community Engagement Classification in 2014. The Chemistry Department has no formal connection to service learning to date, however, our current understanding of STEM Identity concepts have led to discussions to use and expand SCA outreach events to count towards the Civic Engagement graduation requirements. The TCNJ Bonner Center, the campus outreach center, is actively working with the Chemistry faculty to develop service learning activities for the General Chemistry laboratory, which have the demonstrated ability to help retain students by allowing them to understand the greater good of their discipline beyond the traditional classroom (*25–28*). We are currently working with the Bonner Center and Isles, a local community based organization focused on local environmental issues and sustainable communities, to develop assays to analyze Trenton area soil samples for different heavy metals, including lead and cadmium. Students in the General Chemistry laboratories would be participating in a larger, more meaningful research and service project. This would provide students with a sense of purpose and to see the value of chemistry to society, again adding to the STEM Identity.

IV.D. Alumni Events

Recent Department strategic planning aims to increase the interactions with our alumni in order to better track student career trajectories, expand the endowment donor base, and develop the mentoring and networking activities with our current students. The SCA takes an active role in fostering a diverse array of alumni mentors from various career paths using their chemistry degree, including law, industry, teaching, graduate school, health careers, and government work. This wide array of role models gives students a chance to broaden their view of what they can *become* in the wide-ranging field of chemistry, and to talk to alumni about their specific career trajectories and how they have overcome obstacles and adapted to changes in their plans. These conversations support students developing their social capital through building networks, cultural capital about different kinds of and paths toward chemistry careers, while simultaneously facilitating student transition from a chemistry major to a professional identity. Faculty typically organize the alumni events with SCA guiding the career path suggestions and advertising for students to attend.

We found that alumni are excited to come back to TCNJ to talk about their experiences with current students and to visit once again with their previous faculty members. We have several formal and informal events. The informal events include a Homecoming cookout during the fall football game. We work with the campus Alumni Affairs who provides a tent and space in the alumni tailgating area at no cost to SCA. SCA and the faculty provide tables, lawn chairs, and food (~$200) for all current students and visiting alumni. We have found young alumni that have graduated within the last five years are the most frequent visitors with a few long term graduates that already come annually for another longerstanding group, typically a Greek organization or varsity/club sports group. SCA also hosts a holiday alumni gathering the week between Christmas and New Years. We have had a wide range of alumni that have come back, but fewer current students due to timing. The people who attend this gathering are targeted for our formal events, seminars, and graduation speakers. Many of our alumni are willing to come back to TCNJ to give talks on their current work and this adds to the scholarly culture of the Department. In 2015, we started an alumni colloquium series, in which an alum from the Department gives a talk for the current students and attends a lunch with students. We give the speaker some TCNJ memorabilia as a gift (~$25) for coming back and sharing experiences. Our final event is held in conjunction with the College's annual alumni weekend. We host a "Welcome back to the Department" social event with snacks and chemistry demos. We anticipate as we incorporate alumni into the culture of our department, we will have more consistent long term alumni that will come to our events to serve as rich mentors with a variety of careers.

For the last two years, we invited an alum to give deliver the graduation speech at our Departmental graduation ceremony, and our past speakers include a lawyer and an inner city teacher. The first speaker was dedicated to non-traditional careers with a chemistry degree and how he still used his scientific thought processes in law. The inner city school teacher spoke about social justice issues and helping those who are less privileged. We wanted to find speakers that give our students a broader perspective about the role of the chemistry major in a variety of careers. In addition, emphasis on giving back to the community can be especially important as part of a science identity for underrepresented students and women (9). The Alumni interaction piece strongly supports the *becoming* component of the Chemistry Major Identity, which in turn aids the transition to a pre-professional identity.

IV.E. Fundraising

In order to host all of these events, a bountiful budget must be available as we spend about $1000-$1200 per year. Internally, SCA raises funds via ACS study guide, clothing, and personal protective equipment (PPE) sales. The Department uses the standardized General Chemistry and Organic Chemistry ACS exams as course final exams. We organize the ACS study guide sale and make a $3 profit on each book. One of our most successful fundraising endeavors is the annual t-shirt, sweatpant, or sweatshirt sale. A TCNJ chemistry major provides a design for the clothing and these items are purchased by students and faculty. These shirts once again reinforce the tight, family-like community of the Chemistry Department and provides a proud sign of belonging and *being* recognized as a Chemistry Major. SCA sponsors a goggle sale for the first year students and SCA members come to the CHE099 class to sell them. The purchase of safety glasses not only promotes a culture of safety in our Department, but, as most chemistry majors own their PPE, having their own safety glasses serves as a marker by which majors can be recognized by faculty, each other, and other students.

In addition to these fundraisers, SCA also applies for grants from the ACS. Students have learned from their advisors, who have active research funding, to search for opportunities and requests for external funding to support larger projects. The SCA has been quite successful with receiving grants both from TCNJ and the national ACS. From TCNJ, students found that the College Union Board offers significant funding for programming activities to conduct off campus excursions. With the help of the advisor, the executive board wrote a successful grant to host a trip to a paintball arena by describing the community and team building that occurs with a competitive sport. From the national ACS, the students have been successful in obtaining the Student Inter-Chapter Relations grants. As with all grant funding, we examine the reviews carefully to improve our grants and programming. In our first experience with the grant, SCA's ideas were too broad with three separate field trips. The project was still funded with the recommendation that we reduce the field trip to a single trip; the students chose the Chemical Heritage Foundation (CHF) in Philadelphia to interact with students from the local Philadelphia colleges and universities, followed by a dinner funded by the grant. The funding has seeded an annual trip to the CHF. Subsequent grants have requested funds to go to the Crayola Factory, The New Jersey State Police Forensics Lab, and the Liberty Science Center. These activities required the students to make connections and collaborate with other student member chapters at area colleges and universities, a universal skill that is required to be a science professional.

The successful grant writing fits marvelously into the *being, becoming,* and *doing* domains of all pre-professional chemistry identities. Students understand that funding is needed to do their chemistry activities, which is no different than a professional in industry or academia. They develop budgets and plan to execute their projects. After completion, they can reflect, assess, and target the next funding source.

V. Conclusions

The SCA has been a labor of love by the most recent advisors, Dr. Abby O'Connor, Dr. Benny Chan, and Dr. David Hunt. We have cultivated the club to grow and achieve national recognition. We have developed a collaboration with Dr. Lynn Gazley to understand the club and its impact through a different lens to allow the Department to use the club to further the pedagogical goals of the Chemistry Department, allowing us to prepare and educate amazing chemistry majors that can tackle a variety of problems. The Department hopes to use the club's activities to further promote non-traditional, but important professional content including service-learning and entrepreneurship. One of our recent graduates remarked in an anonymous exit survey:

"I had a wonderful experience in the chemistry department at TCNJ. After interacting with other majors on campus, I can confidently say that I believe ours is one of the best. I think this not just because we are organized and have an interactive SCA club, but also because all of the full time faculty genuinely love their job and care about their students. Coming in as a freshman, I did not have any idea what it meant to be a chemist and now, I know how to problem solve, lead groups and think creatively. I would not be where I am today without the guidance of each faculty and staff member in the chemistry department."

This student illustrates the growth in identities that occur through our department that we hope all students undergo. SCA is a key component to a larger culture of learning by faculty, students, and staff at TCNJ.

Another increasing popular research based strategy that we have not examined is facilitating and formalizing Communities of Practice within and outside of the classroom (29). The Communities of Practice framework complements the identity and cultural capital components we already use; the attention to identity trajectories and inclusion of new members is especially helpful. One such example could be the Greek chemistry organization Alpha Chi Sigma where students form a community that can navigate social and academic domains of the Chemistry Major Identity. Another example is where a course instructor organizes and facilitates students to meet outside of the classroom to solve larger problems. After our study of the social science aspects, we believe faculty should look beyond their courses and curricula and examine the chemistry learning that occurs within student life.

Although we have not formally obtained Institutional Review Board (IRB) approval, the club could be rigorously studied to gain additional information on the social schemas at work at TCNJ. Intertersectionality and our developing framework of the *becoming, being,* and *doing* aspects of a chemistry major and pre-professional identity is flexible and will guide diversity and anti-bias programming for our faculty and students to make TCNJ Chemistry a more inclusive environment. We believe other departments may be able to use this research-based framework to request and allocate additional resources from their institutions to prioritize their student member groups to further pedagogical goals beyond course content to develop their majors' identity. The activities conducted by student member groups nationwide are not only fun, but the activities develop a Chemistry Major Identity in ways the traditional classrooms do not.

Acknowledgments

The authors would like to thank the people and groups who have supported the programs and activities of the SCA: although the NSF did not directly fund this work, we thank Dr. Donald Lovett, Dr. Sudhir Nayak, and Dr. Lynn Bradley for the useful discussions of the best practices on student success strategies from the PERSIST programs (*4*, *5*) and the NSF for facilitating collaborations between social scientists and STEM educators (*6*); Dr. Lynn Bradley, Dr. Michelle Bunagan, Dr. Heba Abourahma, and Ms. Valerie Tucci for their development and execution of the CHE099 course; Ms. Joyce Gaiser and Ms. Pam Schmierer for administrative support for fundraising, event planning, demonstration preparation, and purchasing; Ms. Jennifer Sizoo for institutional data analysis; The TCNJ Chemistry Department faculty and Dean Jeffrey Osborn for their support of SCA; the Trenton Local Section of the ACS for financial and leadership support; and finally, but not least, all the amazing TCNJ Chemistry students and alums who executed our vision for SCA to become a nationally recognized student member group.

References

1. The College of New Jersey, National Acclaim. http://tcnj.pages.tcnj.edu/about/national-acclaim/ (accessed June 7, 2016).
2. Western Interstate Commission for Higher Education, Knocking at the College Door. www.wiche.edu/knocking-8th/ (accessed June 7, 2016)
3. The College of New Jersey, Strategic Planning. https://strategic planning.tcnj.edu (accessed June 7, 2016).
4. Lovett, D. L.; Osborn, J. M.; Bradley, L. M.; Chan, B. C.; Nayak, S. B. PERSIST in Biology and Chemistry (Program to Enhance Retention of Students In Science Trajectories in Biology and Chemistry), National Science Foundation Grant #0807107.
5. Chan, B. C.; Lovett, D. L.; Bradley, L. M.; Nayak, S. B. PERSIST 2.0 in Biology and Chemistry (Program to Enhance Retention of Students In Science Trajectories in Biology and Chemistry), National Science Foundation Grant #1259762.
6. Nayak, S. B.; Pulimood, S. M.; Chan, B. C.; Gazley, J. L; van der Sandt, S. FIRSTS (Foundation for Increasing and Retaining STEM Students) Program: A bridge program to study the sociological development of science identities, National Science Foundation Grant #1525109.
7. Fox, M. F.; Sonnert, G.; Nikiforova, I. Successful Programs for Undergraduate Women in Science and Engineering: "Adapting" Versus "Adopting" the Institutional Environment. *Res. Higher Educ.* **2009**, *50*, 333–353.
8. Malone, K. R.; Barbarino, G. Narrations of Race in Stem Research Settings: Identity Formation and Its Discontents. *Sci. Educ.* **2009**, *93*, 485–510.
9. Carlone, H. B.; Johnson, A. Understanding the Science Experiences of Successful Women of Color: Science Identity as an Analytic Lens. *J. Res. Sci. Teach.* **2007**, *44*, 1187–1218.

10. Jackson, P. A.; Seiler, G. Science identity trajectories of latecomers to science in college. *J. Res. Sci. Teach.* **2013**, *50*, 826–857.
11. Tan, E.; Barton, A. C. Unpacking science for all through the lens of identities-in-practice: The stories of Amelia and Ginny. *Cult. Stud. Sci. Educ.* **2008**, *3*, 43–71.
12. Holland, D.; Lachicotte, W.; Skinner, D.; Cain, C. *Identity and Agency in Cultural Worlds*; Harvard University Press: Cambridge, MA, 1998.
13. Collins, P. H. *Black Feminist Thought: Knowledge, Consciousness, and the Politics of Empowerment.* Routledge: New York, NY, 1991.
14. Steele, C. M. *Whistling Vivaldi*; W.W. Norton & Company: New York, NY, 2010.
15. Chang, M. J.; Eagan, M. K.; Lin, M. H.; Hurtado, S. Considering the Impact of Racial Stigmas and Science Identity: Persistence among Biomedical and Behavioral Science Aspirants. *J. Higher Educ.* **2011**, *82*, 564–596.
16. Tonso, K. Student Engineers and Engineer Identity: Campus Engineer Identities as Figured World. *Cult. Stud. Sci. Educ.* **2006**, *1*, 273–307.
17. Gazley, J. L.; Remich, R.; Naffziger-Hirsch, M. E.; Keller, J. L.; Campbell, P. B.; McGee, R. Beyond Preparation: Identity, Cultural Capital, and Readiness for Graduate School in the Biomedical Sciences. *J. Res. Sci. Teach.* **2014**, *51*, 1021–1048.
18. Bourdieau, P. The Forms of Capital. *Handbook of Theory and Research for the Sociology of Education*; Greenwood: Westport, CT, 1986; pp 241–258.
19. Ovink, S.; Veazey, B. More Than "Getting Us Through:" a Case Study in Cultural Capital Enrichment of Underrepresented Minority Undergraduates. *Res. Higher Educ.* **2011**, *52*, 370–394.
20. Erickson, L. D.; McDonald, S.; Elder, G. H. 2009. Informal Mentors and Education: Complementary or Compensatory Resources? *Soc. Educ.* **2009**, *82*, 344–367.
21. Burt, R. S. *Brokerage and Closure: An Introduction to Social Capital*; Oxford University Press: Oxford, U.K., 2005.
22. Kim, D.-H.; Schneider, B. Social Capital in Action: Alignment of Parental Support in Adolescents' Transition to Postsecondary Education. *Social Forces* **2005**, *84*, 1183.
23. Tucci, V. K.; O'Connor, A. R; Bradley, L. M. A Three-Year Chemistry Seminar Program Focusing on Career Development Skills. *J. Chem. Educ.* **2014**, *91*, 2071–2077.
24. Cameron, C.; Lee, H. Y.; Anderson, C.; Byars-Winston, A.; Baldwin, C. D.; Chang, S. The role of scientific communication skills in trainees' intention to pursue biomedical research careers: A social cognitive analysis. *CBE Life Sci. Educ.* **2015**, *14*, 1–12.
25. Zlotkowski, E. Introduction. In *Service-Learning and the First-Year Experience: Preparing Students for Personal Success and Civic Responsibility* (Monograph 34); Zlotkowski, E., Ed.; University of South Carolina, National Resource Center for the First-Year Experience and Students in Transition: Columbia, SC, 2002; p x.

26. Esson, J. M.; Stevens-Truss, R.; Thomas, A. Service-Learning in Introductory Chemistry: Supplementing Chemistry Curriculum in Elementary Schools. *J. Chem. Educ.* **2005**, *82*, 1168.
27. Sutheimer, S. Strategies to Simplify Service-Learning Efforts in Chemistry. *J. Chem. Educ.* **2008**, *85*, 231–233.
28. Donaghy, K. J.; Saxton, K. J. Service Learning Track in General Chemistry: Giving Students a Choice. *J. Chem. Educ.* **2012**, *89*, 1378–1383.
29. Wenger, E.; McDermott, R.; Snyder, W. M. *Cultivating Communities of Practice*; Harvard Business Publishing: Boston, MA, 2002.

Chapter 9

The Xavier University of Louisiana Student ACS Chapter: An Organization Guided by a University Mission

Michael R. Adams*

Department of Chemistry, Xavier University of Louisiana, 1 Drexel Drive,
New Orleans, Louisiana 70125, United States
*E-mail: mradams@xula.edu

The Xavier University of Louisiana Student Chapter of the American Chemical Society is one of the larger and most visible student organizations on the Xavier campus. The chapter has been recognized for excellence by both the ACS and the University on numerous occasions. Based on the university mission to promote a more just and humane society, the organization has a strong commitment to community service and much of the programming the club offers can be related to this mission. Solid programs of leadership development and peer mentoring allow for the continued success of the group.

Introduction

Xavier University of Louisiana is unique as the only institution of higher education in the United States that is both Catholic and historically Black. According to the mission statement, "The ultimate purpose of the university is to contribute to the promotion of a more just and humane society by preparing its students to assume roles of leadership and service in a global society". All members of the university community, including faculty, staff, students, and alumni, are familiar with this mission. Much of what we do, both inside and out of the classroom, is driven by this mission statement. The student chapter of the American Chemical Society (aka "Chem Club") is no exception. While their array of activities each year is extensive and varied, their success over the years can be tied to their dedication to this mission, and living the mission through a desire to serve the community and develop leaders. The events and activities they sponsor serve both the campus community and the larger New Orleans community.

As stated in the chapter's constitution, the purpose of the organization is to promote scholarship and awareness in Chemistry and Applied Chemical Sciences. It is also clearly indicated in the stated purpose that this is accomplished, in part, through community service and strong mentoring relationships. Community service and mentoring, both internally and to external groups, are core components of the work of the group and allow student members to remain true to the University mission.

Recruitment and Mentoring of Members

New members are inculcated to the club's culture of service through aggressive recruitment and mentoring. With guidance from the club's two faculty advisors and eight officers, these efforts have led to a stable annual membership of 80 or more students over the past several years. Entering freshmen indicating an interest in chemistry as a major receive a letter from club officers midway through the summer prior to their matriculation. The focus of this letter is to welcome these students to the department and to introduce many of the resources available to them. More specifically for the organization, though, this letter introduces new students to what the club has to offer. Most importantly, recipients of this letter are informed of a peer mentoring program, Chem Connections, through which they can be paired with an upperclass mentor. The benefits of this mentoring relationship are described within the letter and students are encouraged to join the program prior to the start of the fall semester. The program was initiated by club officers several years ago and all aspects of the program, including identification of mentors and pairing with new students, are controlled by the club. The following description of the program is included in the letter:

> "Congratulations on your decision to attend Xavier University of Louisiana. We at Xavier have confidence that your undergraduate experience will be yet another success, and the ACS Chemistry Club has the Chem Connections program to help you make the experience better. Chem Connections is a mentoring program through the ACS Chemistry Club that is designed to provide Freshmen Chemistry majors with a mentor, an upperclassman to help students like you become acclimated to Xavier University. Mentors in the program include sophomores, juniors, and seniors who are majoring in chemistry or biology. A number of first and second year Doctor of Pharmacy students also serve as mentors. Your mentor will be available to listen and to answer any questions you may have about whatever you want to know about Xavier or college life in general. There are many things that upperclassmen wish someone had told them earlier in their academic career to make life a little easier, and now you have that opportunity. We look forward to your arrival in the fall. Remember, one of the keys to enjoying your undergraduate experience is to get involved."

New students are invited to the first club meeting to learn more about this program. In fact, many are paired with their mentor prior to this meeting if they respond through email to request a mentor. Additional advertising of the mentoring program and the first club meeting takes place through classroom visitations during the first week of the fall semester. The first meeting always takes place during the first week of the fall semester and attendance regularly exceeds 100. At that meeting, a final invitation to secure a mentor is offered to all new students, including transfers and those who might not be majoring in chemistry.

While the individual benefits of the mentoring program for new students seem obvious, the payoff for the club is enormous. It is no doubt the most successful component of the recruitment plan for new members, and along with visitations to General Chemistry classrooms and participation in the annual activities fair during New Student Orientation, returning club members have ample opportunity to interact with new students at a very early point in their time on campus. This early intervention is critical, as new students will no doubt find many other opportunities for co-curricular activities being presented to them throughout their first semester on campus. Even though some students participating in the program do not remain terribly active in the club for the duration of the year, these mentoring relationships often persist. Even beyond the first year, some mentor-mentee relationships continue and students have stated that they remain active in the club owing partially to this mentoring experience.

The pool of candidates for the mentoring program is fairly large, with upwards of 200 new chemistry majors enrolling annually. However, the vast majority of these students are enrolled in the two-year Chemistry-Prepharmacy program (a reorganization resulted in the Prepharmacy Program moving into the Chemistry Department in 2005) and many of them will find interest in other organizations such as the Student National Pharmacy Association (SNPhA) and the Prepharmacy Student Association (PPSA). Rather than trying to compete with these groups, the Chem Club has reached out to them by offering to cosponsor events or advertise the activities of these organizations. For example, recently the members of Kappa Psi, a fraternity for pharmacy students, have partnered with the Chem Club in providing mentors for new students. The mentors they have identified are current Doctor of Pharmacy students and this has helped the Chem Club tremendously in pairing new Chemistry-Prepharmacy students with appropriate mentors.

Because of the large number of students requesting mentors, the pool of upperclass students willing to serve in this capacity must be large. While some years it has been necessary to place more than one new student with each mentor, this is not the desired approach. A strong effort is made to retain members in order to serve as mentors, and club officers will frequently reach out to upperclassmen who may not be club members but who have been identified as potential talented mentors. Upperclass students will often step up to help out and, in fact, they decide to become more involved in other Chem Club activities.

The role of a peer mentor can be critical to the success of a new student and frequently there are upperclass students who lack confidence that they can be effective mentors (1, 2). In order to address this, all mentors are provided with training in the form of a peer mentoring guidebook. Much of the content of this guidebook is based on published best practices for peer mentoring and new mentors are asked to review this material before entering their mentor-mentee relationship (3). The six-page guidebook includes a number of tips for being an effective peer mentor, but the most significant message is that there is no single, defined set of skills that a student must possess in order to be a mentor. The suggestions offered in the guidebook have allowed students to use their own judgment to decide how best to develop as effective mentors.

The Chem Connections program has been in place now for well over 10 years. Recently, the University has dedicated new funding to all academic divisions and departments in order to develop more formalized mentoring programs for new students. The Chemistry Department program uses a group peer mentoring approach, but many of the upperclass mentors are drawn from the Chem Club mentoring program. The club has combined efforts with the department to offer activities, both new and old, aimed at retention of first-year students. One such event that has been traditionally sponsored by the club for years is a preregistration social, a session where first-year students can ask the advice of upperclassmen while building their course schedules for the following semester. While it is made clear to students that this is not meant to replace the consultation they should have with their academic advisors, the advisors are well aware that students will typically seek input from upperclassmen. These sessions provide an opportunity for this to take place in a somewhat more structured way and with upperclass students in whom the advisors have confidence to provide appropriate advice. The club and the department are continuing to develop additional joint mentoring activities.

Executive Board and Membership

A key to the continued success of the group is an executive board composed of experienced, dedicated, and knowledgeable student leaders. Without being overly restrictive, the organization has taken care to ensure that this happens through the structuring of the executive board and the guidelines for voting membership. In addition to traditional positions of President, Vice President, Secretary, and Treasurer, the board includes four additional members. Two positions are for community service co-chairs. These board members are empowered to plan all community service events and to identify community partners for these activities. Two additional board members are those elected as Ms. Chemistry and Mr. Chemistry. These two members work together to plan social events and to publicize all club events. While these and additional duties are described in the club's constitution, an added benefit to the organization is that these members participate in biweekly executive board meetings. Oftentimes these students eventually choose to run for higher offices in subsequent years. Thus, the club benefits from a strong pipeline of experienced leaders. It is not uncommon to have higher positions in the organization filled by students who already have two years of service on the board.

Over the years, the frequency of general body meetings has varied from monthly to biweekly. However, executive board meetings have followed a fairly consistent biweekly schedule. Because the number of board members is large and the duties are well defined, these meetings tend to run smoothly with a clear agenda each time. Rarely have we had officers who did not understand or were unwilling to fulfill their roles.

Elections for executive board positions take place in late spring so that new officers can begin their work over the summer. The board always has its first formal meeting before the start of fall classes in order to set in motion plans for the academic year. One immediate goal is to begin planning a full week of activities for National Chemistry Week, an effort that is led by the Vice President. Several of the activities that are described in later sections of this chapter are often components of the club's NCW program.

As mentioned earlier, a large number of students typically attend the opening general body meeting each year. As with many organizations, though, interest and attendance wane a bit as students become busy with academic commitments. In order to combat this, written in the constitution are guidelines for two levels of membership. In order to be regular members, students simply need to sign in at any meeting or join through our campus on-line membership system (OrgSync). All regular members are required to pays dues and to attend general body meetings somewhat regularly. However, in order to run for office and to vote in officer elections, students must be "Active Members". Active membership requires students to pay dues, participate in fundraisers and two community service events per semester, attend one club-sponsored seminar or panel discussion each semester, attend two social events per semester, and attend at least half of the scheduled general body meetings. These guidelines help to ensure that students voting and running for office have some level of dedication to the group, and elections rarely become simple popularity contests. An additional benefit to active members is that financial support for travel to national meetings is only available to active members who are also ACS student members. Dues of $5 per semester (or $5 per year for those who join ACS as student members) are used to support a variety of club activities, including outreach activities and food for meetings and seminars. With the exception of travel to national meetings and the voting restrictions described above, all club activities are open to all members.

The club has also introduced a committee system to encourage member participation. The second general body meeting is conducted as a miniretreat where all members in attendance are asked to join one of several committees, e.g., fundraising, social events, National Chemistry Week, community service. These committees meet for about 30 minutes to discuss ideas for the academic year and then share these with the larger group. The committees will then meet with varying frequency throughout the year to execute these plans.

The club's recruiting efforts, along with these membership guidelines, have helped to ensure a consistently healthy membership. The total undergraduate enrollment at Xavier is approximately 2200 students, and the ACS Chem Club is one of the largest student organizations on campus. Each year approximately 80 - 100 (or more) students meet some level of membership in the club, with 50 % - 75 % meeting the "Active Membership" and ACS requirements. While the vast majority are Chemistry majors, significant numbers of Biology majors, Psychology majors, and others participate each year.

Community Service and Outreach

Each year the club plans an ambitious agenda of community service activities. Led by the two community service co-chairs, these events are both numerous and varied. Helpful to this effort are signature events that the club includes in their schedule each year. These events provide the basis from which to build an extensive program. They also serve to lessen the negative impact of having plans for new events that occasionally do not come to fruition. Two of these events are Super Science Saturday and the ACS Louisiana Local Section Student Research Poster session. Also included annually is an extensive outreach effort in local schools.

Super Science Saturday

Super Science Saturday in an event that is planned in conjunction with the ACS Louisiana Local Section and the Louisiana Children's Museum. Hosted at the museum, this event is scheduled each year to coincide as closely as possible with National Chemistry Week. The local section invites all colleges and universities in the region to participate in the event and asks each participating school to plan at least one hands-on science activity for children ranging in age from about 4 to 10 years old. While most participating groups plan one or two activities, the Xavier Student Chapter regularly plans four. They normally will include one or two of the same activities every year, with Alka Seltzer Rockets being one of the more popular activities. They also will experiment with one or two new ideas each year. Ideas for these new activities often come from the Chem Demo Exchange at ACS national meetings.

With several hundred children attending this all-day event, there is ample opportunity for a large number of club members to participate. Each year, anywhere from 20 to 30 members assist in planning and executing the activities. All members attending the event are required to participate in a training session so that they are prepared to assist the children to complete the activities in the safest way possible. Additionally, the students learn how to discuss the science behind the activities at a level appropriate for the children attending.

This event is one of the more popular activities for newer members of the club, and many of them subsequently become involved in some of the organization's more extensive outreach activities that are described below. The schoolchildren are always excited to work with college students, but both groups benefit. It has been reported that college students who participate in such outreach activities develop improved communication skills and better understanding of basic science concepts (*4*).

ACS Louisiana Local Section Student Research Poster Session

Each fall semester, usually in early October, the Xavier student chapter plans and hosts an event at which any undergraduate or graduate student in the local region can present the results of their research. Financial support (mainly for refreshments) is provided by the ACS Louisiana Local Section, but the event is planned entirely by Xavier students. While many students will present work that is on-going, the timing of the event is specifically designed to allow students who have recently completed summer research experiences to share their findings. Additionally, because the event is somewhat casual, student participants can often use this as a "first attempt" at presenting their work to other scientists. Frequently these students will continue their work throughout the academic year and present more complete results at the spring ACS national meeting or other scientific conferences.

When this event was first envisioned several years ago, the goal was to have it rotate among several colleges and universities in the greater New Orleans area. However, it has remained on the Xavier campus for 14 years now, primarily because the local section leaders have been satisfied with the efforts of the Xavier students. The benefits to Xavier students have been numerous. The convenience of having the event on our own campus has facilitated the participation of Xavier research students. Each year approximately 20 to 30 submissions are received, and on a regular basis 50 % or more of these are from Xavier students. Perhaps more importantly, though, many younger students on our campus are able to attend the event and begin to explore their interest in research. Total attendance for the event normally ranges from about 60 to 100 students, faculty, and other local section members. Thus, an additional benefit to student attendees is the opportunity to network with local professionals.

Outreach in Local Schools

Opportunities for outreach activities with local schoolchildren are abundant in the Greater New Orleans area, and the need for such activities is significant. It is not uncommon for middle and high schools in the area to offer little or no lab experience for their students. Xavier's identity as an historically Black university and its location in New Orleans instill in chapter members a strong desire to work with historically underrepresented groups. Recent data show that approximately 90 % of students enrolled in New Orleans public schools are African-American, and the city has seen an increasing enrollment of Hispanic students since 2005 (5, 6). It is through these efforts to work with local schools that club members truly live the University mission.

Although projects vary from year to year, efforts to offer outreach programs in local schools are significant. At least one extensive, year-long project is completed each year, but several projects of shorter duration are usually planned, too. Many of these projects have as a broad goal filling the gap in hands-on science experience for local schoolchildren. The significant impact that such activities can have for underfunded schools has been documented (7).

Oftentimes, funding for these outreach activities has come in the form of ACS Student Chapter grants. Over the years Community Interaction Grants (CIG, formerly CISA) and Innovative Activity Grants (IAG) have provided seed funding for a variety of projects through which club members have worked with local schoolchildren of all ages. All projects have been student planned and faculty advisors have been largely hands-off when it comes to the writing of grants. Student leaders have gained valuable grant writing experience and have learned how to develop project budgets and write progress reports. Even when not required, a strong effort has been made to raise matching funds for all of these grants.

Projects that have been funded in recent years include:

- Wow Chemistry Wednesdays: Students prepared hands-on activities and traveled to local high schools to teach a class every other Wednesday. Partnerships were developed with three local schools for this project, with a specific focus on providing laboratory experience for students who otherwise would have none.
- Chemistry in a Box: Students prepared kits of simple equipment and supplies that could be used by elementary and middle school students for hands-on experiments and activities. Xavier students would visit various schools to demonstrate the activities and to meet with schoolchildren. The box would then be left with a teacher who could provide hands-on opportunities for additional students.
- Learning Science Matters: Focusing primarily on schoolchildren in grades K-6, hands-on activities having an emphasis on materials science were developed. Interactions with students took place twice per month during after school programs at local elementary schools.
- Elements for Those in Elementary: This project involved only one partner school, Benjamin Franklin Elementary, but several visits to 4th and 5th grade classrooms occurred. Because a regular schedule of visits was planned, the elementary schoolchildren knew to expect their Xavier mentors and they looked forward to the activities that were planned. Multiple activities with a common theme (e.g., for "Phases of Matter", students made Gack and Alka Seltzer Rockets) were planned for each visit, and a strong emphasis was placed on helping the children develop explanations for what was observed.

The extensive program of community service activities has been enhanced by forging strong relationships with other students and professionals who share a desire to serve students in the New Orleans area. Projects such as Wow Science Wednesdays have evolved through a partnership with Dr. Mehnaaz Ali, who began organizing an independent outreach effort in local schools several years ago. Other projects (e.g., "Learning Science Matters") have received both monetary and staff support from Xavier's Partnerships for Research and Education in Materials (PREM) program, directed by Dr. Lamartine Meda.

Several club members have also participated as volunteers in the immensely popular STEM Saturdays sponsored by the STEM NOLA program under the direction of Dr. Calvin Mackie. Xavier's faculty liaison to STEM NOLA, Dr. Florastina Payton-Stewart, has organized groups of Xavier volunteers for these activities since the inception of the program in 2013. The Chem Club now works directly with Dr. Payton-Stewart to help identify and train volunteers to facilitate hands-on STEM activities with local schoolchildren on one Saturday each month at various locations throughout the city. An added benefit for club members is that they are able to network with local chemists, engineers, and other professionals at these events.

Other Service Projects

While the bulk of service work of the club centers on outreach in local schools, several additional activities, both on-campus and off-campus, are planned each year. While some involve science, many others are more general. Some of the more recent activities are described below:

- Ronald McDonald House: Club members purchase food and prepare a meal for those who are temporarily residing at the Ronald McDonald House while family members are in New Orleans receiving medical treatment.
- Kick-it For Cancer: Members of the club organized a campus kick ball tournament as a fundraiser. Teams representing several campus organizations participated in the event.
- UNCF Walk for Education: Several members participate in this fundraising event each year and often walk together as a group.
- Agrowtopia: Members work with others to maintain the Agrowtopia urban farm on Xavier's campus. Agrowtopia provides fresh, healthy produce to those in the surrounding area.
- Safety Shower Testing: The Chem Club works with our department safety committee to test all safety showers and eye wash stations in teaching and research labs.
- Summer Chemistry Research Luncheons: Each year the Chem Club sponsors one in a series of informal luncheons designed to allow students who are conducting research in the Chemistry Department during the summer to present and discuss their work with their peers and faculty.

One of the goals of an aggressive approach to identifying and organizing service activities is to allow all members of the organization to have opportunities for participation. Scheduling is always a challenge, but by planning both weekend and weekday events, as well as on-campus and off-campus activities, most members can easily participate in at least a couple of activities each semester.

Professional Development Opportunities

Because the group has such diverse membership, they often find it challenging to provide sufficient professional development activities to satisfy all members. The officers want students to have as many such opportunities as possible and have made efforts to address this in a number of ways.

Scientific Meeting Attendance

Each year the group has a fairly large number of members attend the spring ACS national meeting. While many of them will present the results of their research, a significant number attend purely for professional development opportunities. In planning schedules for all students at the national meeting, an effort is made to have as many as possible attend events that will benefit the club. Thus, while the more senior members are participating in such events as the Graduate School Reality Check or various technical sessions, new members of the club are encouraged to attend sessions focused on outreach activities in order to bring new ideas back to the club. Members with previous national meeting experience often take the lead to assist others in constructing personal schedules for the conference.

Funding for travel is always a challenge, but the club has been creative in securing sufficient financial support so that individual members have only minimal out-of-pocket expenses. The group regularly receives ACS travel grants and they have been very successful in securing significant funding from our Student Government Association. The SGA requires a clear explanation of the benefits to the students and the University of meeting attendance, and this has prompted club representatives to think beyond simply the opportunity to travel. As a result, the club has historically had among the highest budget amount approvals of all clubs on campus for the past several years. Club members also work with other students and faculty who have grant funding for conference attendance. They are able to keep costs reasonable by sharing hotel rooms, ground transportation, etc., with students who have such funding. This cooperation is facilitated by the club officers and advisors doing all of the planning for travel for all attendees, including those students and lab technicians (recent graduates) who might not be club members. Finally, only those students who meet "active membership" requirements can receive travel funding from the organization. Thus, first year students are generally not included, but the incentive of being able to travel during their second year keeps them involved in the club.

All ACS student members in the club are part of the Louisiana Local Section, and efforts are made to attend meetings and other activities sponsored by the local section as frequently as possible. Club officers will often agree to use money in the club's treasury to cover some or all of the cost of attendance for these functions (e.g., meal costs for dinner talks). Xavier faculty are quite active in the local section and most members of the section are aware of the success of the student chapter. In fact, in 2015 the local section officers decided to create a position for an undergraduate on the executive board and a Xavier student was elected as the first person to fill this position.

The high visibility for the student chapter at different levels of the ACS has resulted in the club frequently being asked to help with national and regional meeting planning. In 2003, 2008, and 2013, the chapter hosted the traditional Sunday evening undergraduate social during the spring ACS national meeting. The approach the officers have used each time is to form a steering committee early in the fall and to seek partner organizations from other local institutions (e.g.,

Loyola University and Dillard University) to assist in planning. Faculty advisors have purposefully played only a small role in this planning so that the student leaders are able to gain valuable experience in executing such a large event. The students are given freedom to secure entertainment, plan activities, and design the agenda for the evening, and part of this planning requires that they remain within the budget provided by ACS. An added benefit for the club is that several first-year students who would not normally be exposed to a national meeting are able to participate as members of the planning committee.

The club has also been involved in planning the undergraduate program for the two most recent Joint Southeast/Southwest Regional meetings in New Orleans. The SERMACS/SWRM joint meetings are quite large by regional meeting standards and undergraduate attendance is fairly high. Planning for the most recent of these meetings began about 18 months in advance with identification of a project manager (a club member who was not an officer), securing of the support of the Loyola student chapter to function as a partner organization, and development of a clear program with a unifying theme. A proposal for this program was funded by ACS and students were given full control to develop and execute their plan. This included selection of speakers for technical sessions (e.g., "The Art of Chemistry" and "Organic Made Simple"), invitations to student chapters to share ideas through a "Community Outreach Exchange", and organizing a happy hour social event with the local YCC group. Again, advisors remained largely silent in designing the program and a large number of students were able to develop their leadership and organizing skills. The program was well-received and the chapter will almost certainly seek to design a program for the next such meeting in New Orleans in 2020.

Invited Speakers and Panel Discussions

The Chemistry Department at Xavier has a robust seminar program and Chem Club members are always encouraged to attend these weekly research talks. In order to supplement this program, the club tends to design its own series of internal and external speakers around career opportunities and talks on the lighter side of chemistry. Topics have included the Chemistry of Chocolate (to coincide with Valentine's Day), Chemistry in Television and Movies, and Chemistry in Art. The club provides a light meal (beyond just snacks) for all attendees and advertises these "dinner talks" around campus. In order to increase attendance and defray costs, other clubs such as Phi Lambda Upsilon (the Chemistry Honor Society) and the Biology Club are invited to cosponsor appropriate events.

A number of panel discussions are also organized, always involving dinner. A particular effort is made to include a pharmacy career panel each year in order to satisfy the interests of a significant number of members. While the club can always rely on Xavier faculty to fill in as speakers and panelists, the goal each year is to identify professionals from outside the university for these events.

Other Activities

Fundraising

In order to offer the wide array of activities already described, a fairly significant level of funding is necessary. Thus, the group must supplement the sources of funding already described by carrying out numerous fundraisers throughout the year. While traditional fundraisers such as car washes, raffles, and candy sales are common, the most significant on-going fundraiser was developed as a result of working with faculty in the department to determine ways in which the club could raise funds that would benefit the department and its students. Enrollment in general chemistry classes is about 500 students each semester and a required item for all students is a laminated data sheet that includes a periodic table and specific data (e.g., K_{sp} values, thermodynamic data, etc.) to be used for quizzes and exams. For many years this item was sold through the Chemistry Stockroom as part of a package that included a course handbook. The club asked for and received permission to prepare and sell the laminated data sheets and the result was that they could sell it for less than it was costing students previously and still make a healthy profit. The club members purchase the supplies and print and laminate several hundred of these each semester. The items are sold by club members in class during the first week each semester, with the added benefit that this gives them additional time to publicize the club and recruit new members. There is general agreement that this has been a winning relationship for all involved.

Social Events

Students will often express frustration that there are not enough social events on campus, so the Chem Club tries to fill this gap by planning a number of fun and sometimes educational activities each semester. Recurring events include a finals study jam (music, snacks, and games) at the end of each semester, game nights once or twice per year, and off-campus movie or bowling nights. The group is always eager to try out new events and some recent activities have been a Tie Dye Your Labcoat event and a Making of Chocolate (hands-on) session. The duties of Mr. and Ms. Chemistry include planning these events and efforts are always made to find cosponsors for larger campus social activities.

Conclusion

The Xavier Chem Club has a long record of success and quality programming. The foundation they have built has brought them to the point of regularly offering at least one or two unique activities each week, such that all members have a menu from which to choose in order to remain active. Much of what they have been able to accomplish can be attributed to:

- A strong program of leadership development that creates a continual supply of effective leaders.

- Members who are dedicated to service and the mission of the University.
- Students who support the success of their peers through mentoring.

Members have developed an extensive program of service aligned with the University mission to contribute to the promotion of a more just and humane society. The primary focus of this service work continues to be the underserved K-12 students in the New Orleans area.

Also in line with the University mission, former club members have used skills developed while serving on the executive board to move into Student Government Association leadership positions and similar positions in other campus organizations. Graduates have served in leadership positions in such organizations as the Student National Medical Association and the National Organization for the Professional Development of Black Chemists and Chemical Engineers.

For all of their hard work, the Xavier Student Chapter of the American Chemical Society has been recognized as an Outstanding Chapter by the ACS for ten years between 2004-2015. The Xavier Center for Student Leadership and Service honored them with the Outstanding Community Service Award in 2013, and they were recognized as the campus Organization of the Year in 2014.

Acknowledgments

Dr. Ann Privett served as an advisor for the organization for many years and was the guiding mentor for the vast majority of the group's outreach activities during that time. More recently, Dr. Candace Lawrence has joined the advising team. Other Xavier Chemistry faculty who have provided guidance and assistance to the club include Drs. Mehnaaz Ali, Florastina Payton-Stewart, Lamartine Meda, and Terry Watt. Funding has been provided by the Xavier Student Government Association, the American Chemical Society Undergraduate Programs Office (through CIG, IAG, IYC, and travel grants), the Xavier PREM Program, and various Xavier faculty who have made significant donations.

Finally, the dedication of hundreds of members and dozens of officers must be noted. Specifically, the following former presidents are to be thanked for their multiple positive contributions: Kennedi Crosby, Lydiah Mensah, LeAnn Love, Shirmir Branch, Ashley Matthew, Julian McKnight, Marian Gray, Trevonne Walford, Shari Johnson, Tony Davis, Nichole Guillory, and Ayanna Jackson.

References

1. Rodger, S.; Tremblay, P. The Effects of a Peer Mentoring Program on Academic Success Among First Year University Students. *Can. J. Higher Educ.* **2003**, *33*, 1–18.
2. Hill, R.; Reddy, P. Undergraduate Peer Mentoring: an investigation into processes, activities and outcomes. *Psychol. Learn. Teach.* **2007**, *6*, 98–103.

3. Omatsu, G. The Power of Peer Mentoring: Peer Mentoring Resource Booklet. https://www.csun.edu/sites/default/files/EOP_Peer_Mentor_ Booklet.pdf (accessed June 22, 2016).
4. Rao, S.; Shamah, D.; Collay, R. Meaningful Involvement of Science Undergraduates in K-12 Outreach. *NSTA WebNews Digest, J. Coll. Sci. Teach.*; May 1, 2007. http://www.nsta.org/publications/news/story.aspx?id= 53844 (accessed June 22, 2016).
5. Department of Education: Louisiana Believes. https://www.louisiana believes.com/resources/about-us/10-years-after-hurricane-katrina (accessed June 22, 2016).
6. The State of Public Education in New Orleans, 2103 Report. http://www. coweninstitute.com/wp-content/uploads/2013/07/2013_SPENO_Final.pdf (accessed June 22, 2016).
7. Pluth, M. D.; Boettcher, S. W.; Nazin, G. V.; Greenaway, A. L.; Hartle, M. D. Collaboration and Near-Peer Mentoring as a Platform for Sustainable Science Education Outreach. *J. Chem. Educ.* **2015**, *92*, 625–630.

Chapter 10

Creating a Successful ACS Student Chapter: The PCUPR Model

Lizette Santos*

Pontifical Catholic University of Puerto Rico, 2250 Luis A. Ferré Blvd.
Suite 569, Department of Chemistry Ponce, Puerto Rico 00717-9997
*E-mail: lsantos@pucpr.edu

A successful chapter was born as a Chemistry Club in 1955 at
Pontifical Catholic University of Puerto Rico. On December 12,
1956 the Council of the American Chemical Society recognized
the formation of the Catholic University of Puerto Rico Chapter
of Student Affiliates. Since then, our chapter have received for
several consecutive years' recognition as an outstanding chapter
for our commitment to the organization. Along these years,
several advisors and student members have dedicated all their
compromise to the chapter. Since the beginning of our chapter,
members have been committed to bring the fun of chemistry to
everyone. Letting people know that Chemistry is part of our
daily life is one of our achievements. Recently, our chapter
has been recognized as a Green Chemistry Chapter and for four
consecutive years we have received this award. Commitment,
dedication, compromise, leadership, empathy, and service are
some of our qualities. These qualities have given the chapter
the privilege of celebrating 60 years of success.

Introduction

What is a successful chapter? For 60 years Pontifical Catholic University of
Puerto Rico (PUCPR) has been successful in having an Affiliate Student Chapter
of the American Chemical Society (ACS). It all started when Dr. Fritz Fromm
submitted to the ACS a registration form in order that our university have an ACS
Affiliate Chapter. On December 12, 1956, the Council of the American Chemical
Society recognized the establishment of the Chapter of Student Affiliates of what

was at that time, the Catholic University of Puerto Rico. For several consecutive years since then our chapter has received recognition as an outstanding chapter for our commitment to the organization.

Since the beginning of the Chapter, the people involved, the Institution, and the quality of the students, have been key values for it becoming a successful chapter. As human beings we deal with many changes through the years and we must adjust to these changes. We need to capture first the commitment of a great advisor. Then, we need to attract students who are really in love with chemistry and finally we need the community to get involved. As a chapter, members need to be focused on what they want to achieve and how they can make a difference in reaching out to others. It is important to develop activities that help all audiences understand chemistry in their daily lives. Activities that make the general public become aware of how science, and primarily chemistry, can help better or improve our world.

Advisors

Let's think about the key factors that have made it possible for our chapter to continue being a successful one all these years. First, excellent advisors that have become part of the chapter as facilitators. They allowed students to acquire the enthusiasm and commitment of being part of a prestigious organization. Second, advisors that have shown to have visionary minds and dedication to the work they realized. In 1954, Dr. Fritz Fromm started to work at PCUPR at which time he established the Student Chapter, known as the Chemistry Club at Catholic University of Puerto Rico. The Chapter's first president was the late Dr. Roberto Marquez followed by Professor Julio Rivera. After a few years, Marquez became a PhD chemist in the industrial sector in Puerto Rico and Rivera later became dean of the College of Science at PCUPR. The love and passion of these people for what they believed in was key to the establishment of the Student Chapter at PUCPR.

We lost Dr. Fromm as advisor in September 1960 when he passed away of a heart attack while attending an ACS National Meeting in New York. Professor Nylda Hatton then became the chapter advisor until 1970. She was part of our institution for many years. She went on to become the director of the Chemistry Department and in 1982 she was promoted to Vice-president of Academic Affairs. Unfortunately, she passed away on October 20, 2015.

In 1970, Dr. Rafael Infante became the new advisor. He was a person highly dedicated to chemistry. Under his leadership our chapter kept receiving the outstanding mention award for several years. In 1984, Dr. Lizette Santos took on the Chapters' responsibility and kept its outstanding record. Subsequently, other faculty members became advisors for a short time: Dr. Carmen Velazquez de Collazo, Dr. Jose Escabi, and Dr. Ivette Torres. In 2010, Dr. Lizette Santos returns as the Chapter's advisor. The Chapter has continued to receive outstanding recognition since 2011.

If we take a close look at these outstanding individuals, we reach the conclusion that being part of the Affiliate Chapter as an advisor requires commitment, dedication, passion, love, and enthusiasm. The love for chemistry and the commitment to let others know about the importance of chemistry in our daily lives is a common denominator among these advisors. Along with these qualities, we need to include the love for students, the university, and the community. An advisor needs to be a friendly companion but also a leader capable of showing initiative and discipline. An excellent advisor must love teaching, but also love communicating their knowledge to others.

Students

The second key point is our students. Our institution is proud to have great students. These students are committed to their goals. Students that became members of our chapter showed the leadership and enthusiasm to promote chemistry in our daily life. They became and are involved in community service, outreach programs, and career development activities. Students that decide to be part of our chapter remained affiliated during their academic years at the institution. Becoming a member changes their lives and makes them leaders. Membership helps them learn to communicate their ideas, and become aware of how they can change the world. They continuously have shown great enthusiasm in visiting schools (K-12) to illustrate the importance of chemistry to those students. Our students are leaders, have shown commitment, discipline, and passion for what they do as members of the Chapter.

Advisors and Student Leaders

The third key point is the combination of a great advisor with great students that allowed the organization to develop great activities for every academic year. One of the objectives of our chapter has been to allow K-12 students get involved in the sciences from pre-school through high-school. Every academic year, our chapter visits from eight or ten schools to attract students to science through knowledge of chemistry. Our chapter reaches out to preschoolers, elementary, intermediate, and high schoolers. Schools that are visited are allowed to be involved in the magic of doing chemistry through chemical demonstrations and/or a chemistry show. Hands-on demonstrations are part of these activities allowing kids to be part of this wonderful world.

Another great activity is allowing the general public to become aware of chemistry in our daily life. For four consecutive years, we have had a twelve-hour marathon of chemical demonstrations in Plaza del Caribe Mall, a large shopping center in our community. Demonstrations related to materials, environment, health, acid-bases, and general topics are performed from 9:00 am to 9:00 pm. Every two hours, the theme is changed which allows the public to better understand the importance of chemistry in our daily lives. People of all ages are attracted to our demonstrations. They receive orientation and information about how chemistry is helping them every day.

To help attract students to our institution and allow networking between school teachers and our faculty, we organize two chemistry competitions during our academic year. During the fall semester we invite fifth graders to the

University for a Science Competition and during the spring semester we do the same with high school students for a chemistry competition. These competitions allow students from different schools to interact. Also, while the competitions are taken place, a workshop for their teachers is offered. Household materials are utilized during the workshop so that presentations can be easily performed at their different schools.

Another great activity is the celebration of National Chemistry Week. We initiate the week with an opening of a social activity with music and a presentation of a video from the previous year's activities. Every day of the week we organize an activity which includes conferences, an open-house, and the Chapters' initiation ceremony. During this week we also celebrate the fifth grade competitions and the Plaza del Caribe chemical demonstrations marathon.

Summary

Finally, the chemistry department director and the chemistry faculty have to commit to help the Chapter and support us in all our activities. Our human resources are what makes us undistinguishable. Being a successful chapter is the responsibility of our chapter advisor, students and chemistry faculty that have taken the involvement of such a prestigious organization very seriously. Sharing the wonders of science in our community is memorable to us and to the general public. Our chapter is involved in the biggest challenges of improving people's lives through the transforming power of chemistry. Every day is different! And we can help in making every day different. We are the difference!

Chapter 11

The ACS Student Chapter at the University of the Sciences: Developing a Successful Spark Behind the Science

Courtney Vander Pyl, Joey Harmon, Bryan Figula, Oleg Davydovich, Joseph Duffield, Atithi Patel, Enomfon Ekpo, Aaron Hogan, Vanessa Jones, and Catherine M. Bentzley*

Department of Chemistry & Biochemistry, University of the Sciences, 600 South 43rd Street, Philadelphia, Pennsylvania 19104-4495, United States
*E-mail: c.bentzl@usciences.edu

The ACS Student Chapter at University of the Sciences is an energetic, enthusiastic group of approximately seventy students dedicated to promoting science and sharing their excitement for chemistry within the department, university and local community. Our activities range from mole day to pasta dinners to mock interviews and we strive to develop our members academically, personally and professionally. Recognized as an ACS Outstanding Student Chapter we hope to share ideas on how to promote fellowship, fundraising, service and professional development at a small, science and health care focused university.

1. Introduction

The University of the Sciences Student Chapter is a very interesting group composed of approximately 70-80 students who work hard, study diligently, develop as chemists, perform research, share the science spark and have fun. University of the Sciences (USciences), was founded in 1821 as Philadelphia College of Pharmacy as the first college of pharmacy in North America. Since its inception, USciences has educated students for careers in the pharmaceutical, science, and the health-care industries. USciences is a university focused on natural and physical sciences as wel as health professions. The Department of Chemistry & Biochemistry houses programs in Chemistry, Biochemistry and Pharmaceutical Chemistry, as well as small MS and PhD programs in Chemistry, Biochemistry, Pharmacognosy and Bioinformatics.

There are several aspects about the USciences student chapter that really make it special. As our name indicates we are a strictly, science university, where science is not shunned but embraced. Although we are predominately a small campus of only ~2300 students it is filled with nerd herds, geeks with glassware, and beaker buddies. Since this attitude permeates the campus it allows for some very interesting interactions.

One example of how our campus is dedicated to sciences is the fact that every chemistry, biochemistry and pharmaceutical chemistry major is a member of ACS. This is possible because our Misher College of Arts & Sciences and Department of Chemistry & Biochemistry pays for the student memberships. Our students are very fortunate to have this benefit and they really appreciate it. More importantly our Dean and Chair send the message to fledgling chemists that professional memberships are important, valued and supported.

In sharing our ideas and in trying to uphold the missions of ACS to "*the improvement of the qualifications and usefulness of chemists through high standards of professional ethics, education, and attainments...*" we focus on our strengths and have divided the paper into the following sections: fellowship, fundraising, service and professional development. We hope that these suggestions will be useful to other student chapters.

2. Fellowship

The American Chemical Society Student Chapter at University of the Sciences prides itself on its sense of friendship, family, and fellowship. Since we are a small chapter of only 70 students we can take advantage of this by spearheading new ideas, developing as leaders and growing together as a family. Overall we strive as a chapter to sponsor at least one event per month focused on fellowship. These may range from a game night, to our annual Thanksgiving potluck dinner to our Chemistry in the Community event at the Franklin Institute cosponsored with the Heart to Soul in Action HOSA-USciences. At our "Welcome Back to School" picnic, we usually take a photo to mark the start of the school year with the faculty, staff, and students. We always look so fresh and optimistic! This year, we co-sponsored another special event with EPIC, Expressions of Philadelphia's Indian Culture, which was the Meet, Greet & Eat with our alumna Hetal (Sheladia) Vasavada, who was a Master Chef Season Six contestant. Hetal provided insight during her time on the show and discussed how chemistry helped her cooking skills. More importantly, Hetal brought food in order for everyone to have a little taste of the Master Chef's cooking. However, we have chosen some other of our favorite, annual events below to highlight and discuss in detail.

2.1. Departmental Events

Speed Mentoring

The first year of a student's college experience can be frightening and overwhelming. To help incoming chemistry majors, our student chapter has developed a peer-mentoring program within the chemistry department. At the beginning of the semester we hold a "speed meeting" type event for the upperclassmen to meet the new class of freshmen. At the end of the event, the freshmen choose their mentor. Then throughout the semester the upperclassmen act as confidants and friends to the new chemistry students to help ease their transition into college life. This program helps the new students adjust seamlessly into our chemistry family.

Black Lab Coat Ceremony

The University of the Sciences is known for its various, professional programs such as physical therapy, occupational therapy, and pharmacy. Therefore, each year these programs hold a white coat ceremony to signify the students' transitions from the pre-professional to the professional years. The students undergo this rite of passage by receiving their very own embroidered, white lab coats. The students from the Department of Chemistry and Biochemistry recognized this and created their own lab coat ceremony. However, just to be different, our students choose black lab coats which looked surprisingly good. During an official ceremony their ambition, hard work, dedication, and optimism are noted by having survived the introduction of general chemistry, the rigors of organic chemistry, and the challenges of physical chemistry. After receiving their coat, each senior performs an acid/base titration to represent the change and recited the ACS professional creed. Of course, the ceremony ends with all of the solutions being mixed into one 10 L round bottom to represent their resilient unity and unique cohesiveness as a group of friends, peers, and everlasting colleagues from the University of the Sciences, Department of Chemistry. We are excited to have started this wonderful new tradition. The response to it was very positive and we hope to continue this in the future.

2.2. Campus Wide Events

Mocktails during National Chemistry Week

During National Chemistry week, we host events to involve the entire campus with fun, chemistry related events. It's a perfect way to bring people together while still focusing on chemistry. One of the first events we hold during this week is a night of "Mocktails" and dinner. This is a social event where we use large, NEW lab glassware, including 600mL beakers and 1L graduated cylinders, to create non-alcoholic cocktails. The names of the drinks are equally nerdy and include: Schroedinger's Cocktail, The Brainiac, and Isaac's Apple Cider (Figure 1). We also include food, music, and activities to create a fun atmosphere where everyone can mingle. An event like this allows students to take a break from school and enjoy the company of fellow classmates.

Figure 1. "Mocktails" is a night of food, fun and non-alcoholic drinks like Schroedinger's Cocktail. Photo Courtesy of University of the Sciences.

The Finding of the Moles on October 23

One of our most anticipated events of National Chemistry Week is done on October 23rd, also known as Mole Day. In preparation for this event, about 150 pom-pom moles purchased from the ACS on line store are assembled on index cards with the message "Congratulations-you found a mole!" and a multiple choice question on the back (Figure 2). The night before the event, student chapter members hide them throughout the campus, making sure no one sees them. In the morning, students walk to class in the hopes of finding one of the many, hidden moles. If a mole is found and the question answered correctly the finder receives a prize.

Figure 2. The "Congratulations! You found a Mole" cards prepared to be hidden by the student chapter and found by the university students on Mole Day. Photo Courtesy of University of the Sciences.

Haunted Labs

Every year near Halloween, we host our largest event of the semester, Haunted Labs. As an organization, we come together to decorate one of the floors and labs of our chemistry building as a haunted house. The halls are covered with fake blood and skeletons hang throughout the floor. We also have people walking around and hiding in costumes to scare people as they walk through. The whole campus is invited to attend. During this portion of the event, we perform different Halloween demonstrations to entertain them including a foaming pumpkin, an exploding gummy bear, boo bubbles, and making slime. We love putting on the show each Halloween-a campus favorite!

Miracle Berries on Earth Day

Every Earth Day, our school likes to celebrate with different activities. Our contribution is hosting a table with Miracle Berries which are the fruit of a plant that contains miraculin. They are available at health sites such as Miracle Frooties and are $15.00/ten tablets. Once a berry tablet is dissolved on the tongue, it causes anything sour that makes food that is eaten afterwards taste sweet. On Earth Day, we set up a table with several sour foods such as lemons, sour candy, and even mustard for people to try after eating a Miracle Berry.This simple event allows the whole campus to get involved with our organization while actually "enjoying" some sour lemons.

The Kidnapping of Albert Einstein

The Department of Biological Sciences on our campus has a life size cardboard cutout of Albert Einstein and Charles Darwin. Our ACS student affiliates "may have" kidnapped Einstein on occasion. Threatening photos of him with bottles of acid dangling over his head "may have" been sent if demands were not met. Ransom notes requiring the Biology chair to wear an "I Love Chemistry" t-shirt for Einstein's safe return again "may have" occurred. Of course, all of this is hypothetical and our student chapter is innocent until proven guilty.

2.3. Community Wide Events

Intercollegiate Chemistry Wars

Our student chapter of the American Chemical Society also helped to create the Philadelphia Area, Annual Chemistry Wars. The purpose of this event is to build and strengthen intercollegiate relationships among chemistry departments in the Philadelphia Area through a series of fun, competitive, and educational activities. The program began in Spring, 2014 when our group and the Temple University Student Chapter applied and both received the ACS Student Chapter Inter-Chapter Relations Grant. Chemistry Wars usually occurs on a weekend where all the schools are able to participate. We hold events such as titration relays, electron dodgeball, conversion relays, and chemistry-based trivia questions. The winning team gets to keep the trophy until the next Chemistry Wars. It is always great to socialize, interact and network with other chemistry students around Philadelphia. Since the University of the Sciences is a small school, events such as the above allow us to become more than just a science university. We are able to create bonds as a family. We strive to grow constantly, instilling camaraderie between the members within the department, as well as other departments, clubs on campus, and neighboring universities.

3. Fundraising

Developing a budget and fundraising are necessary factors to have a successful chapter, thus, planning is key! For example, it requires pre-planning of events, requesting funds from our Student Government Association (SGA), finding other funds if necessary while still generating enthusiasm. Obviously thinking ahead, practically an entire year in advance, is essential. Before every semester, our student chapter executive board comes together to determine the amount, and type, of events that will be held throughout the semester. We first begin planning events that will require funding from our student government association (SGA). Typically, we organize at least five SGA funded events for every semester, one of which is considered a big event. In order to ensure that our events are successful, we request the maximum amount of funds for each event but we usually receive minimal funding. As a result, we utilize fundraisers to raise the necessary, remaining money. Therefore, the following section will discuss fundraising events we have found to be successful and fun.

3.1. The Periodic Table, Brownie Bake Sale

Each year during National Chemistry Week, we hold a periodic table, bake sale where members of our chapter bake brownies and create a periodic table by making each one a different element (Figure 3). Each brownie is $0.50 Silver and gold brownies are the most popular so we always have extra brownies to redecorate and ready to fill in the space when these sell. This event is made successful by placing the table in a high traffic area around lunchtime. This event costs $20, lasted about five hours, and raised about $80 dollars. The campus community looks forward to this event, as many enjoy the nice, creative snack in between their classes.

Figure 3. Periodic Table made from Elemental Brownies during National Chemistry Week. Photo Courtesy of University of the Sciences.

3.2. Ping Pong Ball Explosion

When the weather becomes warmer, we like to host an outdoor event called the Ping-Pong Ball Explosion. For the event, we fill an empty soda bottle with liquid nitrogen, cap it, and place it in a trashcan with a large amount of white ping-pong balls and a few colored balls. Prior to the demonstration, we sell students Red Solo cups for $1 and ask the student to put their name on it. The cups are then placed around the trash can and the goal is to have a colored ping-pong ball land in your cup (Figure 4). People will buy cups, place them around, wait for the explosion, hope that a ball lands in their cup and win a prize. This is a campus favorite and always draws attention. Some caveats before doing this experiment. First, be sure to inform campus security of your planned explosion, otherwise mass chaos might ensue. Second, on a windy day it is helpful to put water in the red cups as they are being placed on the lawn. Otherwise, the cups fall over or blow away. Spectators get upset if their cup is upside down. Third, in terms of the experiment, be sure to practice it and have proper supervision and safety. Minor adjustments like making sure the cap is on completely can be worked out during a run through. Lastly, putting some water in the bottom of the trash can helps immensely. Without water, the reaction takes 10-20 minutes. After adding water the explosion will occur in 10 seconds. In the end, the cost of the cups, ping-pong balls, and liquid nitrogen was $30 and the profit of the activity was $60. Obviously this event attracts a great deal of positive attention and gets some reactions going.

Figure 4. The Ping Pong Ball Explosion complete with Red Solo cups to catch the ping pong balls. Photo Courtesy of University of the Sciences.

3.3. Clean Cars for Chemistry: A Waterless Carwash

Each year during the Spring semester, we hold a carwash with a twist to help raise funds. Being chemistry and environmentally conscience, the twist is a "waterless" car wash. We purchase washing concentrate along with the other cleaning supplies including microfiber towels and squirt bottles from Ecotouch.net. The microfibers towels to dry the cars cost $15 for a pack of 12 and the wash concentrate with spray bottles were 1 gallon for $45. This is enough to wash approximately 50 cars. By utilizing a plant extract, we are able to hold a classic fundraiser while saving on average 20-100 gallons of water per car. We work to advertise to the whole campus community, focusing on the faculty and staff. Customers could simply buy a car wash receipt earlier in the week from an ACS member. On the day of the event, they would put the receipt in their window and by the end of the day, their car is cleaned! The total cost of the event is roughly $100 and the profit generated was $170. The event is more directed to the faculty and staff members but was well received by the campus community.

Our student chapter holds about four-five fundraising events a year, which are detailed in the Table 1. Fundraising and proper budgeting play an important role for our student chapter. It allows us to have the funds to pay for networking events, volunteer services and our annual ACS dinner. In order to devise profitable, fundraising events, the leader of the event must be organized and ensure that there is significant attendance to make the event profitable.

148

Table 1. Annual Fundraising events organized University of the Sciences Student Chapter

EVENT	SUPPLIES	TARGET AUDIENCE	HOURS OF WORK	COST	PROFIT
Periodic Table Bake Sale	Brownies, icing, napkins, plates	Students & Faculty	5 Hours	$20	$80
Ping Pong Ball Explosion	Ping pong balls, trash can, liquid nitrogen, red solo cups & 2 liter soda bottle	Students & Faculty	10 minutes	$30	$60
Waterless Car Wash	Cleaning Solution & microfiber towels	Students, Faculty & Staff	6 Hours	$100	$170
Pasta Dinner	Various types of pasta, marinara and alfredo sauce, trays, stands & sternos, serving spoons	Students, Faculty & Staff	3 hours	$115	$500

4. Community Service

Our student chapter enjoys doing service because of the smiles we see on faces when we reach out to help. Each year our student chapter holds a myriad of events, but the events that involve helping others are the most meaningful. We have participated in various, socially aware events such as Pink-A-Thon for breast cancer awareness, the Thomas' Walk supporting the Greater Philadelphia Coalition Against Hunger, dodgeball tournaments supporting the Ronald McDonald House and the Philadelphia Science Festival where we get the opportunity to excite children about science. However, this section will focus on how our chapter creates unique service opportunities, how to fund these activities and how to create some exciting kid-friendly demonstrations.

4.1. Finding Service Opportunities

Since our university is located near central Philadelphia, we have the opportunity to reach out to the local communities and spread the spark of chemistry. This advantage allows us to participate in community events at educational museums, local schools and also community centers. For example, in the Fall of 2014 we invited 40 students from Brooklyn Technical High School to visit our department for the day. This event was filled with tours, demonstrations, and poster presentations in order to show and educate this scientific magnet school about how research is done at the university level. In the spring of 2015, we organized a science day for fifty, local 3rd and 4th graders who spent the day with us making Diet Coke rockets, liquid nitrogen ice cream and genetic bracelets.

The focus of the day is always to show that science is fun. The best part of the day was when we performed the awe inspiring ping pong ball explosion!

Your organization can create these opportunities as well. Our chapter has found success by participating and working with larger organizations around the community. National Organizations like The Boys & Girls Club of America, Big Brothers/Big Sisters, and the scouts can easily be searched to find local chapters. Your organization could also utilize less obvious resources including: the public library, science museums or departmental faculty with school aged children who would love the opportunity to experience science.

Develop a repertoire of experiments and make it easy for your group to take it on the road. Order plenty of supplies in the beginning of the year, develop scripts for each experiment, ask the department for a small lab space or closet to store your science show, then get the word out! We guarantee if you build it, they will come. However, be prepared that some organizations may require background checks and clearances before allowing you to work with children. Consider creating a position within your chapter that is dedicated to helping volunteers manage this paperwork.

4.2. Examples of Kid-Friendly Experiments

In Table 2 we have included a list of our favorite, kid friendly, safe experiments and the topics they cover. One particular experiment that our student chapter enjoys is called the water glove where you begin with placing a coin in a beaker of water. Ask the class to help you think of a way to retrieve the coin without getting your hand wet. The rules are that you cannot use a tool, add heat, or move the beaker. The solution to this dilemma is lycopodium. When placed on the surface, a water glove is formed which allows a hand to dip into the beaker without getting wet. This is an excellent segway into discussing surface tension, bonding energies or polarity. Service is truly important to our ACS chapter since it involves making people happy and educating people on the fun side of science. We take pride in the our community work and hope that we influence people to pursue science.

4.3. Funding For Community Service Events

The National Chapter of the American Chemical Society offers numerous grants that are available to student chapters for community outreach and innovative projects. We specifically utilized the Innovative Activities Grant (IAG) to help prepare our science day for 3rd and 4th graders. There is also a community interaction grant (CIA) which is awarded to help improve the science education for minority children.

Table 2. Kid Friendly, Hands On Sciences Demonstrations

DEMO	*ACTION*	*QUESTIONS*	*DISCUSSION*	*COST*
Diapers & Polymers	Extract sodium polyacrylate by tearing a diaper apart and shaking it in a plastic bag. Add water. Watch the polymer absorbs 100xs its weight.	Why does this happen? What are polymers?	Measuring, Properties and effects of polymers, Consumer Science- which brand is best?	$.50 a student
Boo Bubbles	Mix dry ice, water, dish soap in a closed container. Attach tube to allow gas to escape.	What reaction is taking place? How does dish soap make the bubble?	Chemical Reactions, Phases of Matter	$5.00/ Presentation
Pop Rockets	Place half an alka seltzer into a film canister. Add water. Reattach cap. Drop assembly down a paper towel tube.	Why does this happen? What reaction causes the build up of gas?	Pressure of the gas from the alka seltzer reacting causes the canister to blow. The reaction of citric acid turning into sodium citrate releases CO_2.	$5.00/ 20+ rockets
Changing Voice	Inhale helium from a ballon. Observe that the voice goes up.	Why does this happen? How does a more dense gas react?	When vibrating against a less dense gas the vocal chords move more quickly and the voice goes up.	$2.00 for a mylar balloon

5. Professional Development

There are many aspects to becoming a successful chemist besides balancing equations, titrating acids with bases and successfully performing the Grignard reaction. Additional skills like how to write a research paper, how to conduct yourself on an interview, and determining where to apply to graduate school also need to be developed outside of the laboratory. Since professional development is so relevant in today's electronically, networked world our student chapter works on making students in the chemical sciences prepared for future career opportunities and also helps them address the ethical issues that they will face. In this section we provide some suggestions on how to help your chapter members develop as mature chemists and professionals.

5.1. "Talk Science" as Often as Possible by Presenting

As future scientists, many students do not have experience in presenting research or conveying scientific information to an audience and just this idea alone can seem daunting and scary. However the single best way to overcome your fear of public speaking is practice, practice, practice.

Our University provides many opportunities to practice "talking science". For example all incoming freshmen at the University of the Sciences do group, poster presentations at the annual "Health and Science" poster fair. Group projects are a great way to start. In addition this opportunity allows students to present new information to scientific and nonscientific audiences.

Every spring most universities also sponsor an annual Research Day. (If your university does not, consider starting one. The Council on Undergraduate Research offers excellent resources on how to start research endeavours www.cur.org) Undergraduate and graduate students from all departments are encouraged to present the work that they have accomplished from the past year. This is a great opportunity to present scientific data as well as a chance to network with different faculty and students. Consider checking your local ACS section for research presentation dates. Opportunities to talk about chemistry also exist on regional and national levels. If you are intimidated by a big, national conference, the ACS also organizes regional meetings such as Mid-Atlantic Regional Meetings (MARM) based on where you live. They also hold conference for speciality areas such as ACS the 90th Colloid and Surface Science Symposium or the 20th Annual Green Chemistry and Engineering Conference (GC&E) If you are interested, search ACS regional meetings. If the idea of talking science still intimidates you, try warming up with the International Public Speaking Organizations, Toastmasters which helps develop the public speaking skills of its members in a very systematic, organized manner. (www.toastmasters.org)

Note that generous funding is also available through the ACS program (ACS National Meeting Travel Grant) and several universities frequently offer support through undergraduate awards and travel stipends. If these types of stipends currently do not exist at your institution it does not hurt to ask your chair or dean.

There is a quote that "confidence is the result of having successfully survived risk." So find or even create opportunities at your university. Then go ahead and take that risk because you WILL not only survive but thrive. The excitement and inspiration of discovery awaits!

5.2. Organize a Mock Interview Day

With the help of your department and Career Services Center consider organizing a mock interview day for your chapter. Schedule a day and invite alumni to meet with students for 15-20 minutes in an interview scenario. Consider developing a portfolio to take with you containing your resume, self statements, thank you letters and even business cards (available inexpensively at www.Vistaprint.com) Be sure to ask for feedback as most Career Services departments should have interview forms for this purpose. The benefits of undergoing a mock interview is that you realize it really is not that bad. After our practice interviews juniors at USciences felt more prepared and relaxed for upcoming opportunities.

5.3. Join the American Chemical Society and Other Professional Organizations

Being a part of professional organizations is important for networking, making connections and gaining experience. ACS is especially beneficial for students pursuing a career in the chemistry or biochemistry fields. It increases your networking, builds your resume, and prepares you for your future. ACS has many different resources to help prepare students for the real world. For example, ACS has a list of different chemistry specializations for students to learn about the various chemistry related fields. It is always advantageous to have connections because it could open doors to other opportunities such as scholarships and internship opportunities. For example, if you have always been interested in the chemistry behind brewing beer join the American Society of Brewing Chemists. These organizations can help by giving you more information, connecting you to experts in that field and offering scholarships for aspiring students. These professional organizations can help you find something you are passionate about and help you on your way to success.

Another example is The Younger Chemists Committee (YCC) which is an important part of ACS. The YCC holds monthly meetings that consist of a YCC representative from several universities, employees of the chemical industry, as well as scientists who are interested in getting more involved with younger chemists. In Philadelphia the YCC organizes several events such as a poster session, the annual ACS meetings and chemistry trivia night, which are all open for anyone to attend. Being connected with the YCC is extremely beneficial to our ACS because it allows us to stay connected with the events happening at other universities, and gives us more chances to spread the word about all cool, things chemistry.

5.4. Host a Speaker/Find a Mentor

One problem for undergraduate students is that they have limited exposure to research performed on their campus. However it is important for students to expand their horizons and possibly spark their interest in other thought provoking areas. Try asking faculty members if they know any colleagues that would be willing to present their research to students. This allows students to mingle with professionals in their respected fields, and possibly meet a future mentor. It could also be a great opportunity for students to build connections with the outside, chemistry world. Do not limit the speakers to simply research talks but also include industrial professionals to share their insight on life outside of academia. This type of knowledge is invaluable and can help students make the tough decision about how to begin their chemistry careers.

5.5. Request a GRE or PREP Class

One of the greatest benefits of our department at USciences is that our faculty care and understand the stress of our majors. For example, many of our upperclassmen plan on pursuing higher degrees at renowned graduate or medical school which require above average chemistry GRE or MCAT scores to even be considered for admission. Although the topics on these tests have been taught throughout the years, students felt they needed a refresher to prepare. This idea was presented to the Chair and the following semester a Prep for Careers focusing on GRE and MCAT test taking course was offered. It was nice to work with our faculty on developing a valuable course deemed to help us professionally. Again do not be afraid to initiate ideas and ask your department for help creating opportunities. Chance are they will be supportive and excited by your initiative.

The American Chemical Society Student Affiliate at the University of the Siences works very closely with faculty and staff in the Department of Chemistry and Biochemistry. From hosting professional scientific talks to giving research presentations individually, the students are able to gain much experience and knowledge that will stay with them for a lifetime.

6. Conclusions

The ACS Student Chapter at University of the Sciences thrives on learning chemistry and inspiring that passion in others scientists As an outstanding chapter we hope to continue to educate and motivate more scientists in the future and we enjoyed sharing our ideas in this chapter. It is fun to "share the spark" by hosting speakers, washing cars without water, kidnapping Einstein, exploding ping pongs balls or just drinking some mocktails with friends.

Acknowledgments

We gratefully acknowledge the Department of Chemistry & Biochemistry at University of the Sciences and the Misher College of Arts and Sciences for financial support of our student chapter. We also thank the ACS for supporting us through Intercollegiate Activity Grants and Innovative Activities Grants.

Editors' Biographies

Matthew J. Mio

Matthew Mio is a Professor at the University of Detroit Mercy in the Department of Chemistry and Biochemistry. His research focuses on new transition metal catalyzed cross-coupling reactions. Projects include exploring both the mechanism and synthetic capabilities of these reactions, with particular emphasis on the generation of phenylacetylenes for use in nanoelectronics and supramolecular chemistry. He is also interested in studying the pedagogy of organic chemistry. He has been co-advisor to the Detroit Mercy Chemistry Club (American Chemical Society Student Members) for over 15 years.

Mio holds a B.S. in chemistry from the University of Detroit Mercy and a Ph.D. in organic chemistry from the University of Illinois at Urbana-Champaign. He was awarded a Mellon Fellowship to perform postdoctoral research and teaching at Macalester College (St. Paul, MN). Mio joined Detroit Mercy's faculty in 2002.

Mark A. Benvenuto

Mark Benvenuto is a Professor of Chemistry at the University of Detroit Mercy, in the Department of Chemistry and Biochemistry, and a Fellow of the American Chemical Society. His research thrusts span a wide array of subjects, but include the synthesis of novel materials for water remediation, plus the use of energy dispersive X-ray fluorescence spectroscopy to determine trace elemental compositions of: aquatic and land-based plant matter, food and dietary supplements, and medieval and ancient artifacts. He has also been an advisor to the UDM ACS Student Members Chapter for many years.

Benvenuto received a B.S. in chemistry from the Virginia Military Institute and, after several years in the Army, a PhD. in inorganic chemistry from the University of Virginia. After a postdoctoral fellowship at the Pennsylvania State University, he joined the faculty at the University of Detroit Mercy in 1993.

Indexes

Author Index

Subject Index

Printed in the USA/Agawam, MA
October 25, 2017

661255.001